Synthesis Lectures on Biomedical Engineering

This series consists of concise books on advanced and state-of-the-art topics that span the field of biomedical engineering. Each Lecture covers the fundamental principles in a unified manner, develops underlying concepts needed for sequential material, and progresses to more advanced topics and design. The authors selected to write the Lectures are leading experts on the subject who have extensive background in theory, application, and design. The series is designed to meet the demands of the 21st century technology and the rapid advancements in the all-encompassing field of biomedical engineering.

Arvydas Palevicius · Giedrius Janusas ·
Urte Cigane · Justas Ciganas

Nano/Micro Functional Elements Formation for Bioengineering Applications

Springer

Arvydas Palevicius
Faculty of Mechanical Engineering and Design
Kaunas University of Technology
Kaunas, Lithuania

Giedrius Janusas
Faculty of Mechanical Engineering and Design
Kaunas University of Technology
Kaunas, Lithuania

Urte Cigane
Faculty of Mechanical Engineering and Design
Kaunas University of Technology
Kaunas, Lithuania

Faculty of Business and Technologies
Siauliai State Higher Education Institution
Siauliai, Lithuania

Justas Ciganas
Faculty of Mechanical Engineering and Design
Kaunas University of Technology
Kaunas, Lithuania

Faculty of Business and Technologies
Siauliai State Higher Education Institution
Siauliai, Lithuania

ISSN 1930-0328 ISSN 1930-0336 (electronic)
Synthesis Lectures on Biomedical Engineering
ISBN 978-3-031-81511-9 ISBN 978-3-031-81509-6 (eBook)
https://doi.org/10.1007/978-3-031-81509-6

This Springer imprint is published by the registered company Springer Nature Switzerland AG
The registered company address is: Gewerbestrasse 11, 6330 Cham, Switzerland

Preface

In the last decade, nano/micro functional elements have received a lot of attention from the public and scientists, both from an economic and ecological point of view. Various micro- and nanostructured elements are studied, and applied in different bioengineering fields such as tissue engineering, biomaterials, biomechatronic, biotechnology, bioprocess engineering, etc.

The purpose of this monography is to describe the production of microfunctional and nanofunctional elements, describing scientific achievements and applications in bioengineering.

The scientific monograph is focused on the processes of forming microstructures in different thermoplastics and determining the mechanical properties of thermoplastics. The vibration-based hot stamping process is presented, in which high-frequency vibrations are used to form microstructures in thermoplastics. To optimize the forming process, a forming tool is described, presenting a theoretical and experimental analysis of the tool. Additionally, experimental results of microstructure formation using the described forming tool are presented. Moreover, in this monography, the production of nanostructured membranes is presented, describing in detail the technological processes, equipment, and results obtained in the production of nanoporous anodized aluminum oxide (AAO) membrane and nanopillared chitosan membrane. This monography also presents the main and most relevant areas of application of microfunctional and nanofunctional elements in bioengineering.

This book is useful for practitioners, researchers, and manufacturers in the field of bio-engineering and nano/micro functional element formation. In addition, this monography could be used in advanced courses as supplementary material for doctoral students.

Kaunas, Lithuania
2024

Arvydas Palevicius
Giedrius Janusas
Urte Cigane
Justas Ciganas

Contents

1 Introduction

Bioengineering can be defined as the integration of natural and engineering sciences [1]. Furthermore, bioengineering is recognized as a multidisciplinary science that involves numerous fields, including materials science, chemistry and biochemistry, mechanical and chemical engineering, etc. Consequently, nano-/micro functional elements have been crucial in the fields of bioengineering [2]. Tissue engineering receives particular attention because of its limitless possibilities to enhance human health. Many people experience bone defects each year as a result of trauma, tumors, or bone diseases; therefore, nano/micro functional elements have the potential to open up an abundance of medical applications [3].

Significant advances in both engineering and organ and cell transplantation have led to increased interest in bone tissue engineering [4]. Because organic and inorganic components make up most of the bone, bone tissue engineering could make use of a wide range of nano/micro functional elements and combinations. The materials used most frequently in bone tissue engineering are collagen, polystyrene, acrylates, polyglycolic acid, chitosan, and ceramics (hydroxyapatite or tricalcium phosphate) [5]. For example, carbon nanoparticles have excellent mechanical stability and are widely available commercially [6]. Furthermore, organic materials are often used in bone constructions, such as ultra-high molecular weight polyethylene [7]. Graphene oxide can be used as a nanomembrane for bone tissue engineering due to its high minimum cytotoxicity to live cells and strong biocompatibility [8, 9]. Inorganic materials are another example of nano/microfunctional elements for bone tissue engineering application. Good biocompatibility, little toxicity, a sizable active surface area, and the ability to serve as carriers of the necessary medications or minerals are among the benefits of silica nanoparticles and composites [6]. In

A. Palevicius et al., *Nano/Micro Functional Elements Formation for Bioengineering Applications*, Synthesis Lectures on Biomedical Engineering,
https://doi.org/10.1007/978-3-031-81509-6_1

addition, bone architectures often incorporate inorganic nanomaterials, such as ceramics, pure titanium, and titanium alloys [7].

Skin tissue engineering is an area of significant research because the skin is the largest organ in the human body and provides protection against external pathogens. Therefore, skin anatomy and physiology can be restored using nano/micro functional elements [10]. For example, due to their low toxicity and reduced chronic inflammatory response, natural biomaterials such as chitosan offer several advantages [10]. On the other hand, synthesized nanoparticles offer benefits over natural materials, such as being more affordable and characterized. Metals and oxides are among the inorganic nanomaterials that could be used to heal wounds. Wounds have been treated with silver as a medicinal agent. In other research, several scaffolds were implanted with titanium dioxide (TiO_2) to increase their mechanical strength, a necessary component for the engineering of skin tissue [10]. Other elements, such as carbon nanostructures, for example, are non-toxic, exhibit high mechanical qualities, and can replicate the inherent mechanical strength of bone [11]. The characteristics of silicon carbide nanomembranes include flexibility, durability, specific ion permeability, and chemical stability. An electrospinning method was used to create a mixed nanofibrous membrane with polyvinylpyrrolidone [12]. Research has shown the effectiveness of this membrane and its potential for application in biomedical wound dressings that could accelerate skin wound healing. Furthermore, silicon carbide nanomembranes have been shown to be suitable for flexible implanted devices due to their biocompatibility and the necessary mechanical strengths [13]. Furthermore, research has demonstrated the formation of new hydrogen bonds between the PLLA carboxyl groups and the collagen amide groups. These membranes provide a potentially effective method for making skin wound dressings. Moreover, phase segregation technology has been successfully used to create nanoperforated PLA nanomembranes, and it is suggested that this will lead to major advances in the application of these nanoperforated PLA nanomembranes for tissue regeneration processes [14]. Among organic nanomaterials, graphene oxide serves as a soft nanomembrane with exceptional rigidity and biocompatibility, making it a promising choice for stem cell development [11]. Other studies found that the hybrid nanomembrane produced by electrospinning poly(L-lactide) (PLLA) and skin collagen from *Rana chensinensis* had higher mechanical strength than the natural collagen nanofiber [15].

In addition to tissue engineering, effective drug delivery is gaining more attention these days, and research related to drug delivery systems is still a challenging task. Drug delivery can be achieved with organic nano/micro functional elements [16]. Polymeric nanostructures, such as PEDOT, PEG, and polyglycidyl methacrylate nanotubes, are examples of this type of structure [17]. Nanofiber elements might be other examples. Electrospun poly(N-isopropylacrylamide) (PNIPAM)/gelatin nanofibers demonstrated the release of anticancer medications from the membrane at a specific temperature [18]. Furthermore, since the drug release process takes eight hours, the PCL/shellac/PCL nanofiber sandwich structure membrane met the criteria for use in nighttime skin care treatments on the face and exhibited good mechanical and other characteristics [19]. Furthermore, drug delivery methods frequently use polylactic-co-glycolic acid (PLGA) nanomembranes produced by electrospinning [20]. Research has demonstrated the pain relief properties of the PLGA nano membrane. The nanostructures of poly(hydroxyethyl methacrylate), poly(methacrylic acid) and poly(n-isopropylacrylamide) were created using the chemical vapor deposition technique [21]. Furthermore, by electrospinning PLA/graphene oxide nanofiber membranes with various topologies, it was possible to demonstrate their promising potential as scaffolds in drug delivery systems [22]. There are still several obstacles to overcome before using core sheath nanofibers for drug delivery applications, as noted in the review study [23]. Both these and other nano/micro functional elements have significant impact for application in drug delivery systems because of their necessary pore shape and large surface area.

Numerous novel manufacturing technologies are being developed to produce nano/micro functional elements of higher quality and greater mechanical strength. For this reason, nano/micro functional element technologies in bioengineering fields such as drug delivery systems and tissue engineering still face many challenges and opportunities. As a result, this book introduces new technologies to produce nano/microfunctional elements that could be applied in bioengineering.

The second chapter presents an analysis of the microstructure formation process. The mechanical properties of polypropylene (PP), polyethylene terephthalate glycol (PETG), polyvinyl chloride (PVC), and styrene acrylonitrile (SAN) thermoplastics up to the viscoelastic temperature were determined, and the mechanical properties were also determined. In addition, the vibration-based hot stamping process was simulated using a finite element model.

The third chapter presents an improved magnetostrictive transducer for the formation of microstructures. Theoretical analysis and production of the forming tool was described in detail.

The fourth chapter presents the experimental study of the fabricated magnetostrictive transducer prototype. Two studies were conducted. First, a pilot study of the hot stamping process was performed. Then, a study of the hot stamping process with vibrations was conducted.

The fifth chapter presents the development and analysis of an electrochemical reactor with vibrating functional element for the fabrication of AAO nanoporous membranes. A reactor study was described in detail, where the effects of reactor temperature, mixing process, and vibration were analyzed.

The sixth chapter presents the fabrication of nanoporous AAO membranes using high-frequency during the anodization process. The influence of different frequencies on the pore geometry of the nanoporous AAO membrane was analyzed.

The seventh chapter presents the fabrication of free-standing chitosan membranes using the vibration-assisted solvent casting method. Vibrations were used to control the height of the chitosan membrane nanopillars.

Finally, the eighth chapter presents the application of nano/micro functional elements in different fields of bioengineering. Here, plasmon metal nanostructures for detection and sensing cell-biological particles, microfluidic channels for bio particles manipulation and identification of liquid concentration in the periodic microstructures are presented.

References

1. Dias RR, Zepka LQ, Jacob-Lopes E (2019) Introductory chapter: biotechnology and bioengineering. In: Jacob-Lopes E, Zepka LQ (eds) Biotechnology and bioengineering. https://doi.org/10.5772/intechopen.86380
2. Bhat S, Kumar A (2013) Biomaterials and bioengineering tomorrow's healthcare. Biomatter 3(3):e24717. https://doi.org/10.4161/biom.24717
3. Li X, Liu W, Sun L, Fan Y, Feng Q (2014) The application of inorganic nanomaterials in bone tissue engineering. J Biomater Tissue Eng 4:994–1003. https://doi.org/10.1166/jbt.2014.1253
4. Shadjo N, Hasanzadeh M (2016) Graphene and its nanostructure derivatives for use in bone tissue engineering: recent advances. J Biomed Mater Res Part A 104:1250–1275. https://doi.org/10.1002/jbm.a.35645
5. Jaksic Z, Jaksic O (2020) Biomimetic nanomembranes: an overview. Biomimetics 5(24). https://doi.org/10.3390/biomimetics5020024
6. Eivazzadeh-Keihan R, Chenab K, Taheri-Ledari R, Mosafer J, Hashemi S, Mokhtarzadeh A, Maleki A, Hamblin M (2020) Recent advances in the application of mesoporous silica-based nanomaterials for bone tissue engineering. Mater Sci Eng C 107:110267. https://doi.org/10.1016/j.msec.2019.110267
7. Hill MJ, Qi B, Bayaniahangar R, Araban V, Bakhtiary Z, Doschak MR, Goh B, Shokouhimehr M, Vali H, Presley J, Zadpoor A, Harris M, Abadi P, Mahmoudi M (2019) Nanomaterials for bone tissue regeneration: updates and future perspectives. Nanomedicine 14:2987–3006. https://doi.org/10.2217/nnm-2018-0445
8. Foong LK, Foroughi MM, Mirhosseini AF, Safaei M, Jahani S, Mostafavi M, Ebrahimpoor N, Sharifi M, Varma RS, Khatami M (2020) Applications of nano-materials in diverse dentistry regimes. RSC Adv 10:15430–15460. https://doi.org/10.1039/D0RA00762E
9. Zhang J, Chen H, Zhao M, Liu G, Wu J (2020) 2D nanomaterials for tissue engineering application. Nano Res 13:2019–2034. https://doi.org/10.1007/s12274-020-2835-4
10. Singla R, Abidi S, Dar A, Acharya A (2019) Nanomaterials as potential and versatile platform for next generation tissue engineering applications: nanobiomaterials for tissue engineering applications. J Biomed Mater Res Part B Appl Biomater 107:2433–2449. https://doi.org/10.1002/jbm.b.34327
11. Venkatesan J, Pallela R, Kim S (2014) Applications of carbon nanomaterials in bone tissue engineering. J Biomed Nanotechnol 10:3105–3123. https://doi.org/10.1166/jbn.2014.1969
12. Dai XY, Nie W, Wang YC, Shen Y, Li Y, Gan SJ (2012) Electrospun emodin polyvinylpyrrolidone blended nanofibrous membrane: a novel medicated biomaterial for drug delivery and accelerated wound healing. J Mater Sci Mater Med 23:2709–2716. https://doi.org/10.1007/s10856-012-4728-x
13. Phan HP, Zhong Y, Nguyen TK, Park Y, Dinh T, Song E, Vadivelu RK, Masud MK, Li J, Shiddiky MJA, Dao YY, Rogers JA, Nguyen NT (2019) Long-lived, transferred crystalline silicon carbide nanomembranes for implantable flexible electronics. ACS Nano 13:11572–11581. https://doi.org/10.1021/acsnano.9b05168

14. Puiggalí-Jou A, Medina J, Valle LJ, Alemán C (2016) Nanoperforations in poly(lactic acid) free-standing nanomembranes to promote interactions with cell filopodia. Eur Polym J 75:552–564. https://doi.org/10.1016/j.eurpolymj.2016.01.019
15. Zhang M, Wang J, Xu W, Luan J, Li X, Zhang Y, Dong H (2015) The mechanical property of *Rana chensinensis* skin collagen/poly (L-lactide) fibrous membrane. Mater Lett 139:467–470. https://doi.org/10.1016/j.matlet.2014.10.085
16. Zhang Y, Fang F, Li L, Zhang J (2020) Self-assembled organic nanomaterials for drug delivery, bioimaging and cancer therapy. ACS Biomater Sci Eng 6:4816–4833. https://doi.org/10.1021/acsbiomaterials.0c00883
17. Zhou X, Wang Y, Gong C, Liu B, Wei G (2020) Production, structural design, functional control, and broad applications of carbon nanofiber-based nanomaterials: a comprehensive review. Chem Eng J 402:126189. https://doi.org/10.1016/j.cej.2020.126189
18. Slemming-Adamsen P, Song J, Dong M, Besenbacher F, Chen M (2015) In Situ cross-linked PNIPAM/gelatin nanofibers for thermos-responsive drug release. Macromol Mater Eng 300(12):1226–1231. https://doi.org/10.1002/mame.201500160
19. Ma K, Qiu Y, Fu Y, Ni QQ (2018) Electrospun sandwich configuration nanofibers as transparent membranes for skin care drug delivery systems. J Mater Sci 53:10617–10626. https://doi.org/10.1007/s10853-018-2241-4
20. He Y, Qin L, Fang Y, Dan Z, Shen Y, Tan G, Huang Y, Ma C (2020) Electrospun PLGA nanomembrane: a novel formulation of extended-release bupivacaine delivery reducing postoperative pain. Mater Des 193:108768. https://doi.org/10.1016/j.matdes.2020.108768
21. Armagan E, Ince GO (2015) Coaxial nanotubes of stimuli responsive polymers with tunable release kinetics. Soft Matter 11:8070–8075. https://doi.org/10.1039/c5sm01074h
22. Mao Z, Li J, Huang W, Jiang H, Zimba BL, Chen L, Wan J, Wu Q (2018) Preparation of poly(lactic acid)/graphene oxide nanofiber membranes with different structures by electrospinning for drug delivery. RSC Adv 8:16619–16625. https://doi.org/10.1039/C8RA01565A
23. Pant B, Park M, Park SJ (2019) Drug delivery applications of core-sheath nanofibers prepared by coaxial electrospinning: a review. Pharmaceutics 11:305. https://doi.org/10.3390/pharmaceutics11070305

2 Analysis of the Microstructure Formation Process

Introduction. A revolution is currently taking place in the fields of microstructure creation, development, and application [1]. Microstructure technologies are becoming more relevant, and microstructures are being applied more widely in different fields.

The characteristic dimension of the microstructure is less than 1 mm. Typically, microstructure systems have capillary channels that are connected to external devices such as pumps, valves, sensors, heat sources, etc. [2]. Microsystems have advantages over traditional systems. For example, by performing chemical reactions in microsystems, it is possible to obtain more accurate results using less material in a shorter time interval [3]. In addition, a microsystem-based chemical sensor can detect substance concentrations as low as one part per billion [4]. On the other hand, the use of microsystems in medicine can provide information on changed health indicators [5]. Therefore, microstructures and microsystems provide an opportunity to create more compact, accurate, and efficient devices, and for this reason, they are a promising field [6].

Microstructures have various properties that affect their practical application. One of the most important characteristics of microstructures is geometry, which can be classified by profile, degrees of freedom, and surface roughness [7]. With the possibility of controlling these characteristics, the practical application of microprocesses in various fields can be expanded.

The manufacturing methods and technologies used in the production of microstructures have a significant influence on the profiles of microsystems and their properties, so the manufacturing process must be foreseen and carefully considered in the development process. By choosing the wrong manufacturing technology, the generated geometry can acquire mixing, fluid control, or filtration properties [8]. In addition, improperly selected

A. Palevicius et al., *Nano/Micro Functional Elements Formation for Bioengineering Applications*, Synthesis Lectures on Biomedical Engineering,
https://doi.org/10.1007/978-3-031-81509-6_2

production technology can have a negative impact on the performance and operating principle of the structure.

Microstructures can be formed by various manufacturing methods. To form microstructures, it is necessary to choose production methods taking into account different criteria: the formation of the microstructure in the selected material, the geometry of the structure, the available equipment, whether the production time has a significant influence, etc. Some methods are more suitable for forming structures in laboratories to ensure quality, while other methods are more suitable for mass production because the production method is more sustainable [9]. Microfabrication techniques can be divided into five main categories: photolithography, solidification or stamping processes, etching, deposition processes, and hot stamping. These processes can also be analysed and classified according to the type of material being processed, the dimensions of the structure, or other alternatives.

Microchannel formation in thermoplastics is widely used because of its unique properties. Compared to glass-based structures, plastic structures are more resistant to impact, have higher transparency, lower moulding costs, and the ability to form more complex structures. Due to the low melting point, hot-pressing technology is commonly used in materials such as thermoplastics.

Hot-pressing technology is characterised by high precision and low production costs, so it is often used in mass production. Depending on the type of structure to be formed and the scale of production, there are two forming methods: single-step and roll-to-roll hot stamping. A common one-step forming method consists of two panels. The working material, usually a polymer, is placed on the base plate, and the structure to be formed is attached to the upper sole. The work material is heated to the glass transition temperature using a heating element. The upper pad then presses the work material with the load, changing the geometry of the material. The plates are then separated [10]. In the roll-to-roll method, two rollers are used, one for the microstructure and the other as a support roller. The main roller is heated to such a temperature that the working material reaches the glass transition temperature. A sheet of working material is rolled between the rollers, in which the structure is formed. In this way, a continuous microstructure can be ensured [11]. The methods of the hot stamping process are presented in Fig. 2.1.

Ensuring the levels of the forming process parameters to ensure tool cavity filling is one of the major challenges faced in microstructuring during hot-press technology. The wrong parameter selection leads to a low-quality structure. By choosing the correct production parameters, efforts are made to automate the hot stamping process to produce higher quality results [12].

Further, the formation technologies of nano/micro functional elements and their potential are introduced, as well as basic principles, advantages, and disadvantages. Because different fabrication methods can be used to form nano/micro functional elements [13], a variety of factors must be considered when selecting production techniques to form microstructures, nanostructured membranes/films, including the material's ability to form microstructures, the structure's geometry, the structure, the available equipment, and the

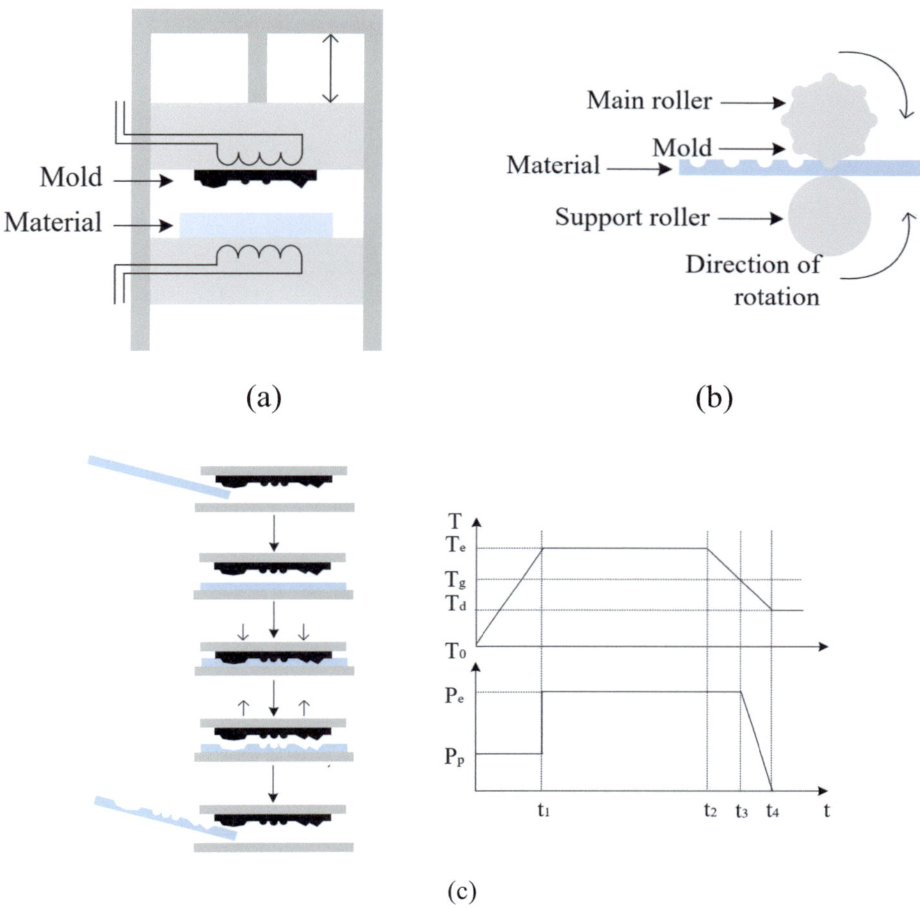

Fig. 2.1 Schematic diagrams of the hot molding process: **a** one-step method; **b** roll-to-roll method; **c** heating stages

fact that production time has a major impact [14–17]. So, the main techniques, such as lithography, micromachining, 3D printing, layer-by-layer deposition, molding, anodization and new technologies of nano/microstructure formation were thoroughly reviewed.

Lithography technology. Many different industries make important use of lithography. These days, the objective is to obtain minimal features with high resolution. This means that greater investigation of lithography technology is needed to obtain better results in the manufacturing of micro-/nano functional elements. Maskless lithography and masked lithography are the two categories of lithography procedures [18]. In masked lithography, patterns are transferred to a base material using a mask. Different designs are produced

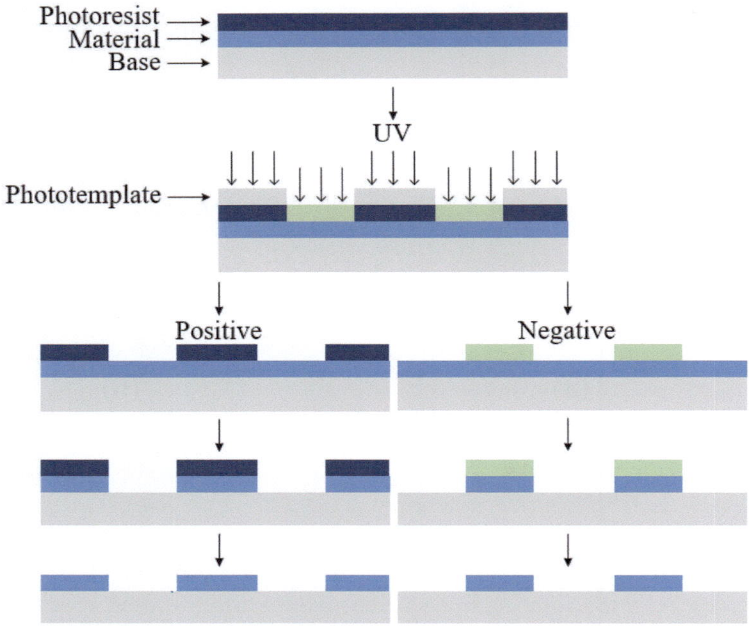

Fig. 2.2 Principle of photolithography

by serial writing in maskless lithography (where the mask is not used). Photolithography (minimum feature size 2–3 μm) and nanoimprint lithography (minimum feature size 6–40 nm) are two examples of masked techniques, while another maskless method is electron beam lithography, which has a minimum feature size of 5 nm [19].

In the photolithography process, there are three main steps: surface exposure, photoresist coating, and surface etching. The principle of the photolithography technique is presented in Fig. 2.2.

In the first stage, the plane is covered with a thin coating of photoresist. The photoresist layer is then flattened by rotating the plate on a machine. Rotating speed, time, and material viscosity all affect the layer's thickness. Dosing from a robotic arm is an option to the levelling procedure [20].

In the second stage, a microstructure is created using ultraviolet (UV) light through the phototemplate. Usually, the UV light source is a special UV lamp, the light flux of which is diffused. Using a light source, the contact photolithography method is applied. High resolution can be obtained using this method, but when there is contact between the material and the photopattern, the product may be damaged or deformed. To avoid material contact, the method has been improved, and the phototemplate is placed at 10–20 μm. Such an enhancement increases the diffraction of light, which is why the maximum

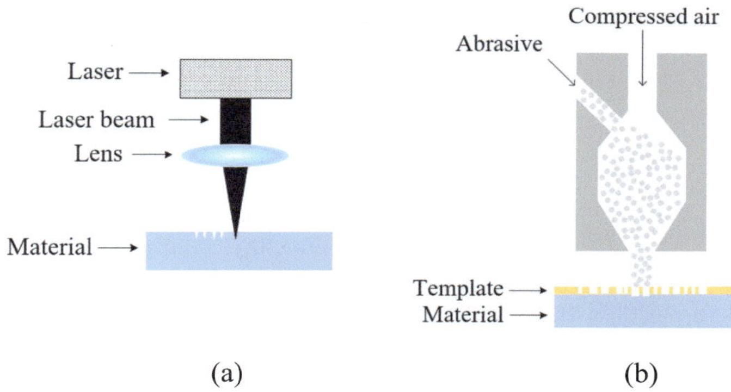

Fig. 2.3 Principles of micromachining: **a** laser processing; **b** abrasive processing

resolution cannot be achieved. To avoid deficiency, it is proposed to use a projection exposure system. This system works in a similar way to a projector and can achieve a resolution of up to 1 μm detail using 300–400 nm wavelength light [21].

The third step is surface etching. During this process, unnecessary material is removed from the surface. When a chemical solvent is chosen to remove the surface of the materials, it is important that the solvent does not contaminate the formed structure. In the case of a complex structure and to remove material residues from deeper parts of the structure, ultrasonic cleaning could be used [22].

The key benefits of lithography are its adaptability and process efficiency. However, the equipment required for lithography makes it a highly expensive procedure.

Micromachining technology. Micromachining can be used to create a variety of functional nano/micro elements in a variety of metals, alloys, semiconductors and polymers [23]. In this study, micromachining is interpreted as one of the latest technologies that has attracted a lot of interest in the creation of nano/micro functional elements [24]. It is a technique of micromachining with a laser. The method of creating the desired design using a laser beam that moves along a specific path is known as laser micromachining [25]. The principles of micromachining are presented in Fig. 2.3.

Another micromachining technique is abrasive etching. Abrasive particles are directed to the desired location using a nozzle in the abrasive etching process. The substance is torn apart by the high-speed impact of the particles. Abrasive particles can range in size from 30 to 100 μm. The abrasive stream can be concentrated and used as a drill or tool to form channels. It is also possible to use an abrasive template that covers areas that do not require machining. In this case, the uncovered areas are treated with an abrasive [26].

For the creation of nano/micro functional elements, micromachining is a very straightforward technique with a wide range of options. However, costly equipment is one of the main drawbacks.

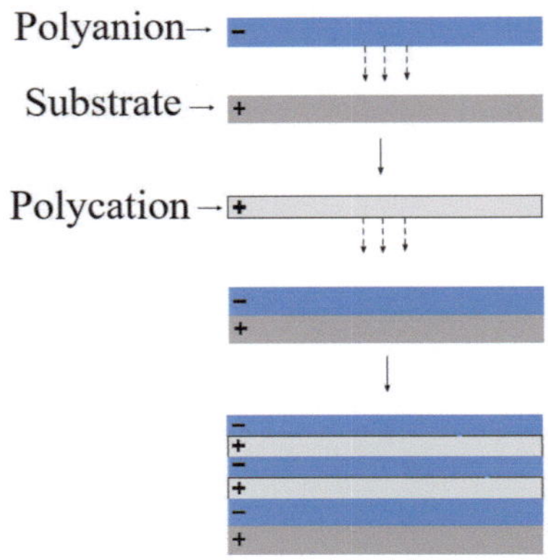

Fig. 2.4 Principle of the layer-by-layer deposition technique

3D printing technology. Recent attention has been paid to the fabrication of nano/micro functional elements using additive manufacturing, or 3D printing, due to its many benefits, including the lack of need for a cleanroom, inexpensive equipment and consumables, easy accessibility, simple design and reprinting processes, support for multiple materials and multiphase printing [27]. However, 3D printing technology has not yet progressed sufficiently to fabricate nano/micro functional elements because of its low resolution. Moreover, 3D printing technology has many different forming technologies, but all these technologies have one fundamental drawback—they are not applicable in mass production, because the structure formation process is relatively long [28].

Layer-by-layer deposition technology. Layer-by-layer deposition is an additional technique for creating nanomembranes that involves the adsorption of various charged molecules. The principle of technology is presented in Fig. 2.4.

As an example, the substrate is submerged in a diluted cationic polyelectrolyte solution. A single monomolecular layer of polyelectrolyte with a thickness of around 1 nm is adsorbed. The wafer is then dried and cleaned. The polish-coated substrate is then submerged in a diluted polyanion solution. Over the layer that was previously deposited, a new monolayer forms. The wafer is then cleaned and dried once more. The required number of layers can be achieved by repeating the operation with other or identical materials. Multilayers with thicknesses ranging from 5 nm to more than 500 nm are produced using the layer-by-layer deposition technique [29].

Molding technology. Using molding technology, plastic is heated within a mold to begin the softening process. The workpiece is molded according to the specifications for use under exterior pressure. Microlens arrays and other tiny, flat optical lenses are good candidates for it. Simple principles underlie the molding equipment that is used to produce small batches and a variety of products. As a result, it shortens the experimental duration and offers financial benefits in the production of experimental supplies and the research of optical plastic materials. The process of hot-pressing technology involves many steps,

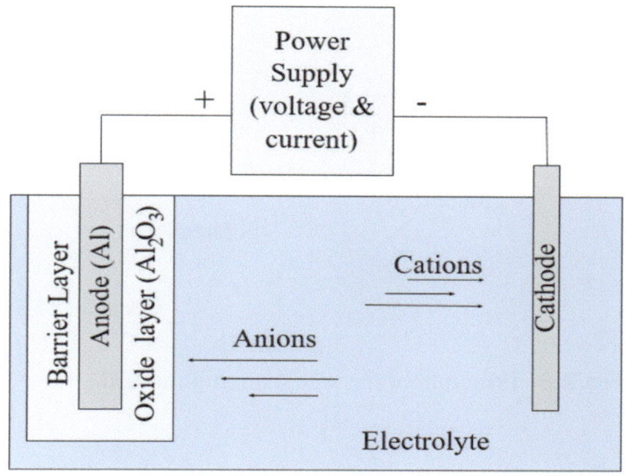

Fig. 2.5 The principle of the anodization process

including heating and preserving heat, hot pressing, slow cooling, and demolding. High efficiency, environmental friendliness, lack of pollution, and suitability for large-scale manufacturing are among its benefits. However, after machining, there is a morphological deformation. This type of technology is good for industrialization as it requires very little equipment, has a fast machining cycle, and produces a lot of work [14].

Anodization technology. The electrolytic passivation technique, known as anodization, is used to increase the thickness of the oxide layer on the surface of metal elements. The principle of the anodization process is presented in Fig. 2.5.

Anodized aluminum oxide (AAO) produced by using aluminum anodization technology is an illustration of the self-ordered electrochemical process. The exceptional qualities of ordered AAO, including its large surface area, hardness, chemical stability, and thermal stability, make it stand out. In the last ten years, there have been a number of applications related to nanostructured AAO elements. The key benefits of anodization include inexpensive manufacturing costs, controllable pore diameters, and well-organized porous structures. As a result, anodization technology is a desirable technique for creating nano/micro functional elements. However, there are disadvantages related to a complex controlled process due to variable parameters and the use of toxic chemicals.

New technologies of nano/microstructure formation. Improved and more flexible techniques have been created. Novel approaches to microstructure formation have their own benefits that are applicable in specific areas. One of the newest technologies, for instance, is the wet stamping method, which localizes electrochemical reactions by using a gel template. The lack of sharp edges in the created structure is one of its drawbacks [30]. The wet stamping method is shown in Fig. 2.6.

The femtosecond laser is a microstructure-forming technology relatively better than the previously mentioned laser technology. Femtosecond laser technology does not leave any

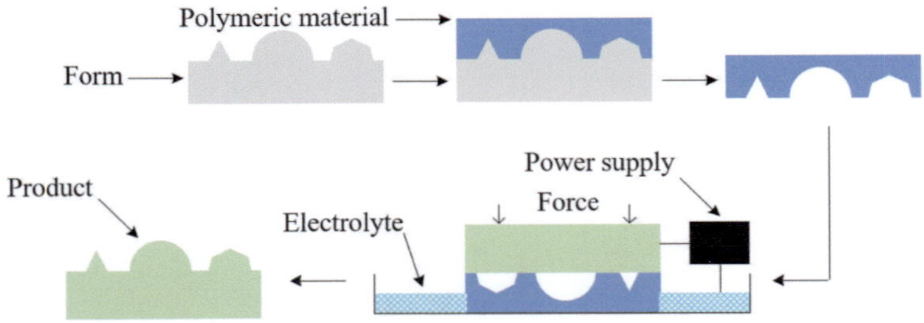

Fig. 2.6 Principle of the wet stamping method

microcracks or residual melting zone, but the technology itself is not yet fully developed [31].

Functional materials. Functional materials are defined as materials that have specific properties or functions. For example, one of the properties could be piezoelectricity, magnetism, or energy storage. Functional materials can also be called materials that have unique properties in their group of materials. For example, among metals, aluminum can be singled out as a functional material, characterized by low density, which makes it possible to use this material in the creation of light structures. When creating microstructures, the functional properties of materials become even more important. The right choice of material can give the required properties to the structure. Different functional materials can be divided into different groups [32], which are silicon, polymers, metals, glass, and others.

Polymers are currently used extensively because of not only their mechanical properties, but also their increasing choice and intensive research in the development, analysis, and application of these materials. Polymers are dielectrics with low mechanical strength and low melting point. One of the main advantages of polymers is their functionality. Polymers can be thermoplastics or reactoplastics. Compared to thermoplastics, reactive plastics are more stable and retain their basic dimensions better under changing environmental conditions. Thermoplastics are used more widely in manufacturing because they can be easily reshaped using simple techniques [33]. The most commonly used thermoplastics are polypropylene (PP), polyethylene terephthalate glycol (PETG), polyvinyl chloride (PVC), and styrene acrylonitrile (SAN).

PP thermoplastic is distinguished by biocompatibility, high operating temperature, low macromolecular adsorption, and nontoxicity [34]. In biology and medicine, PP plastic is used to create reaction tubes. Plastic is also used in medicine in the production of human implants [34].

PETG plastic has very good optical properties, biocompatibility, and the possibility of using this plastic with a 3D printer. These possibilities make it possible to create transparent microsystems of unlimited geometry, which can be used to create fuel cells or micromixers [35]. PETG plastic has a higher glass transition temperature (T_g), for this reason, this plastic can be used as a template for plastics with a lower glass transition temperature [36]. PETG plastic is also applied as a basis for the development of electrochemical microfluidic devices and sensors [37].

PVC has good biocompatibility and is therefore also widely used in the medical device industry to store various fluids (dialysis, blood or blood product solutions) [38]. One of the main disadvantages of PVC polymer is its poor chemical resistance to aromatic solvents, which limits the application of chemical reactors [39]. The microfluidic device developed in this plastic can be applied in cell biology, bioengineering, or chemical analysis of compounds [39].

SAN thermoplastics are characterised by high strength, resistance to chemicals, solvents, and excellent processing properties [40]. However, poor strength and crack resistance lead to fewer applications of this material [41]. Despite the disadvantages, SAN plastic is widely used in areas where its good properties are best revealed. Researchers use SAN plastic in the development of membranes because the plastic has good thermal resistance properties, making it possible to reduce the losses incurred when using the membrane [42].

In an analysis of microstructure formation processes, standard deformation simulations use material properties that describe the behaviour of materials at room temperature. In the hot stamping method, the forming temperature exceeds the glass transition temperature of the material, which is why it is necessary to expand and describe the properties of the materials in a wider temperature range. Therefore, the determination of the mechanical properties of PP, PETG, PVC, and SAN thermoplastics up to the viscoelastic state needs to be analysed. In the case of hot stamping with vibration, additional processes begin to take place in the material, which can be described by a viscoelastic model. The properties of the material are determined by experimental measurement of the stresses, strains, and the angle between the stresses and strains.

2.1 Evaluation of PP, PETG, PVC and SAN Thermoplastics' Mechanical Properties up to the Viscoelastic Temperature

One of the most appropriate ways to develop and study a high-quality microstructure is by using a finite element model. The created model, regardless of the geometry, is easily adapted to the new changed geometry. Such a model allows one to easily determine the necessary parameters for the formation of the structure.

Standard mathematical models analyse the behaviour of a material with elastic properties. This is the behaviour of a material in which a body subjected to deformation can return to its original shape when the load is removed. The opposite process is plasticity, where an object remains deformed and does not return to its original shape when the load is removed.

Fig. 2.7 Tensile testing machine Instron E10000 and control computer

To determine the plastic deformations of thermoplastics, it is necessary to know additional properties of the material. The deformations of the elastic part can be described using Young's modulus, but for plastic deformations, a multilinear isotropic elastic curve is used. Plastic tensile tests are performed to determine this curve.

The behaviour of thermoplastics during tensile tests in the stage of plastic deformation can be characterized and described in several ways [43]. In this work, the Mooney–Rivlin model was used. Tensile tests were performed at room temperature on four thermoplastics. Then, additional temperatures were selected to calculate the multilinear isotropic hardening.

An *Instron E10000* test device (*ElectroPuls E10000T Linear-Torsion*, *Instron*, USA) was used to perform the tensile experiment. The testing equipment consists of regular tensile parts plus an extra heat chamber that was used to generate the temperature range. The equipment of the experiment is presented in Fig. 2.7.

The specimens were formed using a computer numerical control laser machine. ISO 527-2 dog bone specimens were made using a laser cut plate made of PP, PETG, PVC and SAN. The five prepared specimens of each thermoplastic were stretched at different temperatures. All specimens of PP, PETG, and PVC plastics were stretched, but some specimens of SAN plastic, which was characterized by brittleness, broke. After reaching the SAN glass transition temperature, the specimen changed properties and stretched. Photos of the specimens are presented in Fig. 2.8.

In the experiment, the specimens were stretched until they broke, or the maximum limit of the tensile testing machine was reached. A gripper moved at a constant speed of 10 mm/min. To bring the specimens to the required temperature, they were placed and held in a steady-temperature thermal chamber until they reached the set temperature. The results obtained from the thermoplastic deformation and stress experiment are presented in Fig. 2.9.

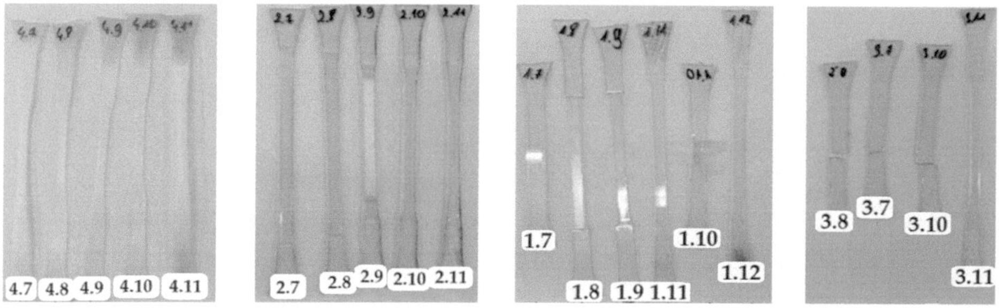

(a) (b) (c) (d)

Fig. 2.8 Specimens after tensile experiment: **a** PP thermoplastic (4.7–4.11); **b** PETG thermoplastic (2.7–2.11); **c** PVC thermoplastic (1.7–1.12); **d** SAN thermoplastic (3.7–3.11)

Using the obtained stress–strain curves, the Young's modulus of each thermoplastic was calculated. Young's modulus was calculated by dividing the stress by the strain. The obtained results showed that Young's modulus decreases with increasing ambient temperature. The results are presented in Table 2.1.

The experimental working temperatures were selected on the basis of theoretical thermal properties. The experimental results showed that the Young's modulus of the thermoplastic PP decreased with increasing temperature from 85 to 125 °C. In an experiment with PETG thermoplastic, at a temperature of 143 °C, the plastic blank suddenly became matte when stretched. The experimental temperature range of SAN plastic was chosen from 60 to 100 °C, and the Young's modulus decreased from 3100 to 1620 MPa.

2.2 Modelling of the Hot Stamping Process Using the Finite Element Method

A finite element model has been created to theoretically analyse how PP, PETG, PVC, and SAN polymers act through the hot-stamping process. The parameters established throughout the experiment have been used by the model to examine hot-stamping technology at the microlevel. The model was made of a matrix and a plastic model. A two-dimensional space was used to illustrate the case.

Using lithography and etching techniques, a nickel matrix was selected for the hot stamping method. Because nickel could be easily regulated throughout the electroplating process and is durable, it was selected as the material for the matrix. Two-dimensional grooves of 2 μm in width, 4 μm in periodicity, and 1 μm in depth composed of the matrix. A simplified unit model was developed for finite element modelling. The model consisted of one groove, half of which was represented for the simulation. The finite element model and the geometrical parameters of the model are presented in Fig. 2.10.

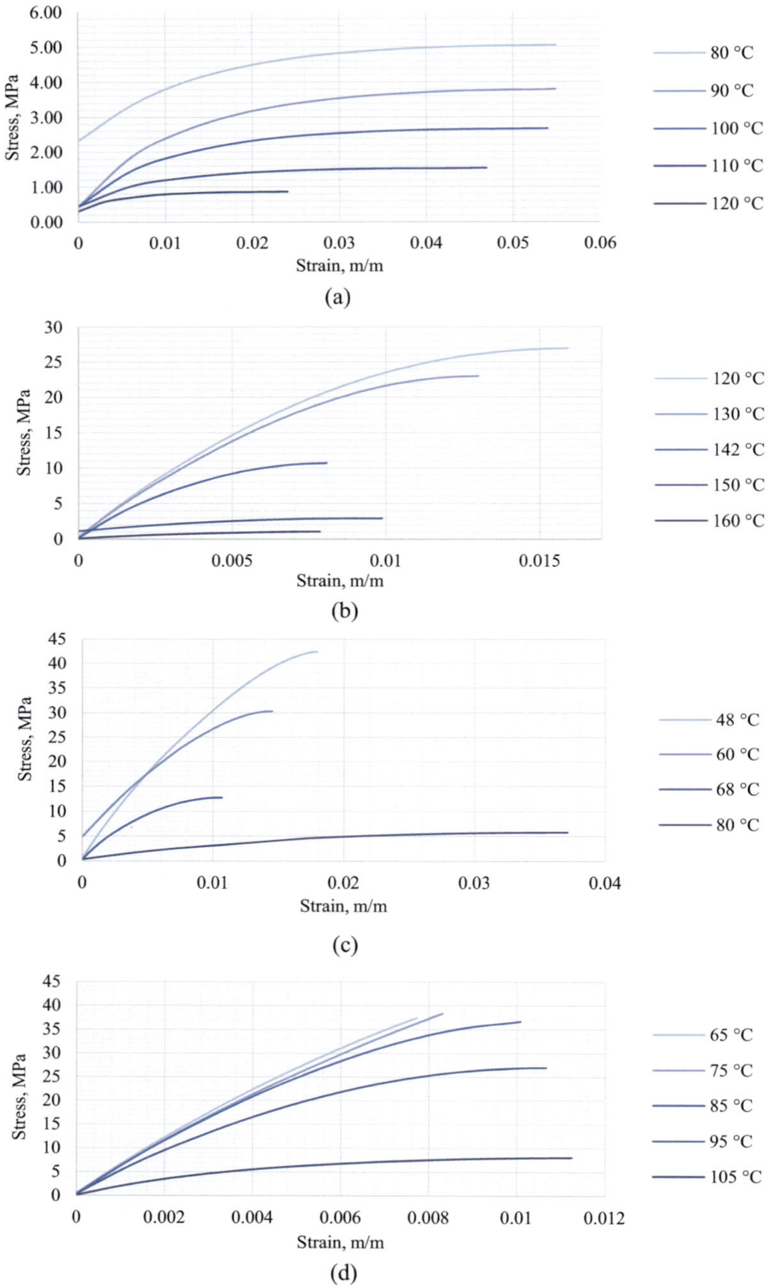

Fig. 2.9 Stress and strain curves of: **a** PP thermoplastic; **b** PETG thermoplastic; **c** PVC thermoplastic; **d** SAN thermoplastic

Table 2.1 Young's modulus at various temperatures

PP	Temperature, °C	85	95	105	115	125
	Young's modulus, MPa	375	282	198	140	100
PETG	Temperature, °C	110	125	140	155	170
	Young's modulus, MPa	3150	3075	2753	300	170
PVC	Temperature, °C	45	60	70	75	85
	Young's modulus, MPa	3050	2717	2005	432	102
SAN	Temperature, °C	60	70	80	90	100
	Young's modulus, MPa	3100	2956	2784	2411	1620

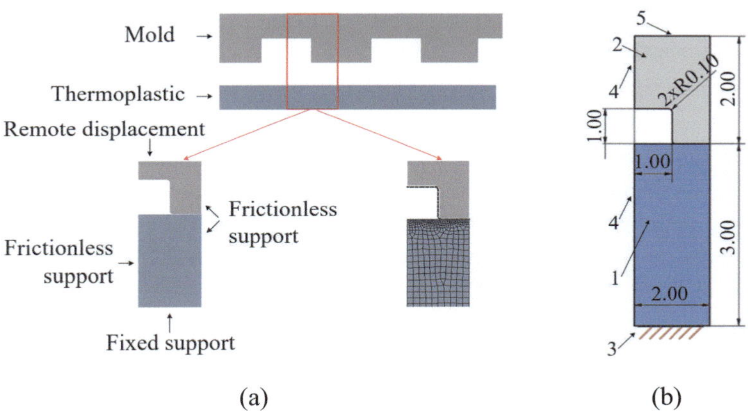

Fig. 2.10 Digital microstructure model: **a** cross-sectional view of the model; **b** a single element model with boundary constraints: 1—substrate, 2—element of matrix, 3—fixed support, 4—symmetry region, 5—an external force

Ansys software (*Ansys*®, Pennsylvania, United States) was chosen for the analysis of the hot stamping process. The Mooney–Rivlin model was used in the analysis, and the experimentally determined properties were interpolated and the stresses and strains at different temperatures were calculated.

A nondeformable mold and a flexible thermoplastic (in stiffness) form the mathematical model that was developed. The parameters were specified as one millimetre in length, and it was described as two-dimensional. Thermoplastic and metal were in frictional contact with a friction coefficient of 0.2. The model was divided into 0.1 μm elements to optimize calculations. In the contact region, the scale has been reduced to 0.025 μm to guarantee precise results. The base was constructed solidly. The integrity of the model was evaluated using a frictionless support on the sides. It was determined that the mold

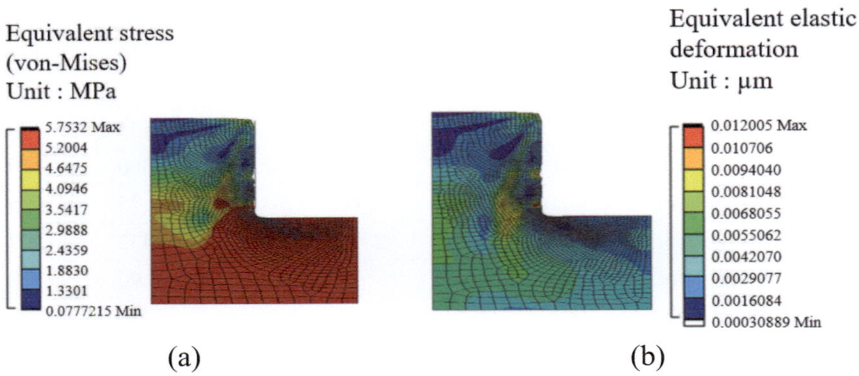

Fig. 2.11 Results of the analysis of the stress and strain distribution: **a** equivalent stress; **b** equivalent elastic deformation

automatically moved up to 2.5 μm toward the plastic before retracting. A nonlinear adaptive domain was used, which recalculated the finite element mesh in the context of a bigger geometry change, for even more precise results. Three output data points were chosen: total strain, equivalent stresses, and reaction force. The images representing the equivalent stress and elastic deformation are presented in Fig. 2.11.

After simulations, the maximum stress, strain, and reaction force results were obtained. The results are presented in Table 2.2.

A 2000 μm^2 area was used to calculate the reaction force. The forming force must be increased according to the dimensions of the matrices to provide the forming force necessary for the microstructure formation process.

The possibility of correctly setting the forming parameters for the most effective forming process has been demonstrated by the simulation results. The results obtained demonstrated that as the formation temperature increases, the response force decreases. Understanding the characteristics of the materials contributes to calculating the reaction force needed for the entire hot stamping process. The optimised model increases the potential of microstructure-forming technologies and helps to produce microstructures of higher quality by making it easy to identify the forming parameters by varying the geometry or the material.

2.3 Evaluation of PP, PETG, PVC, and SAN Thermoplastic Mechanical Properties

This chapter presents a study of the viscoelastic characteristics of PP, PETG, PVC, and SAN polymers at various temperatures and dynamic loads. Viscoelastic properties can be defined in a number of methods. The elastic (storage) modulus (G′), the viscous (loss)

Table 2.2 Results of hot stamping simulations

PP	Temperature, °C	85	95	105	115	125
	Stress, MPa	4.4170	3.2305	2.1025	1.2010	0.4325
	Strain, μm/μm	0.012	0.011	0.011	0.009	0.004
	Reaction force, μN	14,557	10,597	6881	3915	1403
PETG	Temperature, °C	110	125	140	155	170
	Stress, MPa	28.913	26.923	13.532	1.942	0.857
	Strain, μm/μm	0.008	0.007	0.006	0.006	0.004
	Reaction force, μN	94,229	81,217	42,591	8001	2223
PVC	Temperature, °C	45	60	70	75	85
	Stress, MPa	45.912	30.392	10.991	8.590	4.220
	Strain, μm/μm	0.010	0.010	0.006	0.030	0.005
	Reaction force, μN	145,230	100,102	38,929	29,954	17,342
SAN	Temperature, °C	60	70	80	90	100
	Stress, MPa	38.993	38.852	37.553	32.838	18.02
	Strain, μm/μm	0.005	0.008	0.008	0.004	0.005
	Reaction force, μN	127,930	127,543	125,021	105,845	65,050

modulus (G″) and the damping factor were experimentally measured with the aim of determining the viscoelastic properties.

In this study, previously described thermoplastic samples from PP, PETG, PVC and SAN were used. To determine the viscoelastic properties of the materials, dynamic mechanical analysis (DMA) was performed. This study differs from standard stretching because it uses cyclic loading, where the intensity of the load cycling is varied, rather than the magnitude of the load itself. This experiment was performed using the ISO 6721-4 standard, which describes the determination of the properties of polymeric materials at a non-resonant frequency between 0.01 and 100 Hz. The test carried out according to the standard allowed the obtaining of the elasticity values (G′) and the viscosity modulus (G″) and the calculation of the phase between the stress–strain curves (δ).

An *Instron E10000* tensile machine (*ElectroPuls E10000T Linear-Torsion*, *Instron*, USA) was used for the dynamic tensile test. The experiment consisted of active elements of the tensile machine: a passive and active gripper, a thermal chamber, and the specimen. During the analysis, each specimen was held at a specified temperature according to the ISO 6721-1 standard, so the temperature was evenly distributed throughout the specimen. The dynamic stretching experiment is presented in Fig. 2.12.

A maximum dynamic amplitude of 20 N and a constant tension force of 30 N were found for PP plastic. A maximum dynamic amplitude of 100 N and a constant tension

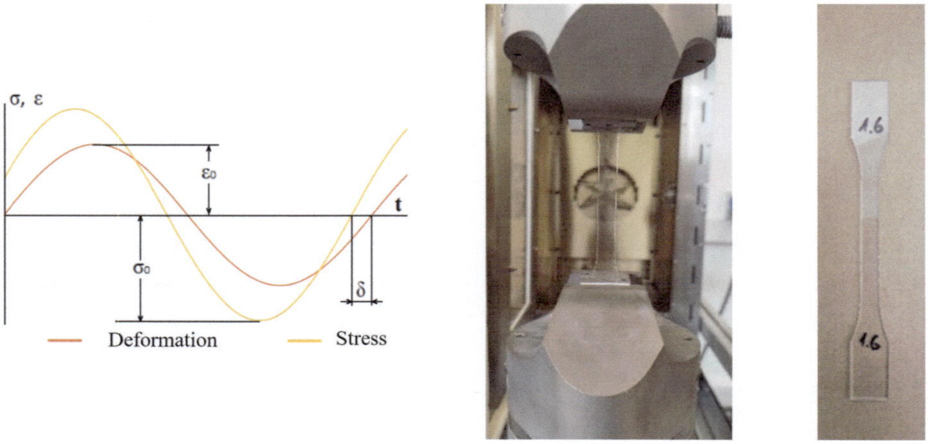

(a) (b) (c)

Fig. 2.12 Dynamic stretching experiment: **a** theoretical graph of phase calculation; **b** specimen placed in the machine; **c** specimen

force of 170 N have been observed for PETG and SAN polymers. For the PVC thermoplastic specimen, a maximum dynamic amplitude of 70 N and a constant tension force of 80 N were determined. At frequencies of 0.1, 0.2, 0.5, 1, 2, 5, 10, 20, 30, and 40 Hz, experiments were conducted. PP plastic was analysed from 40 to 100 °C, PETG plastic from 50 to 140 °C, PVC plastic was analysed from 30 to 66 °C and SAN plastic from 20 to 95 °C. After exceeding the glass transition temperature, the specimens either cracked or stretched to the limits of the testing instrument.

The constants of storage (G′) and modulus of loss (G″) were determined using the formulas:

$$G' = \frac{\sigma_0}{\sigma_0}\cos\delta \tag{2.1}$$

$$G'' = \frac{\sigma_0}{\sigma_0}\sin\delta \tag{2.2}$$

where G'—storage modulus, Pa; G''—loss modulus, Pa; σ—stress, Pa; δ—phase between the stress–strain curves.

The modulus of storage and loss obtained at different temperatures are presented in Figs. 2.13, 2.14, 2.15 and 2.16. These curves present the behaviour of the material up to an excitation frequency of 40 Hz.

The experiment was conducted in the low frequency region (up to 40 Hz) since the *Instron E10000* test equipment (*ElectroPuls E10000T Linear-Torsion*, *Instron*, USA) has a limitation on the excitation frequency that could generate. Since high-frequency excitation

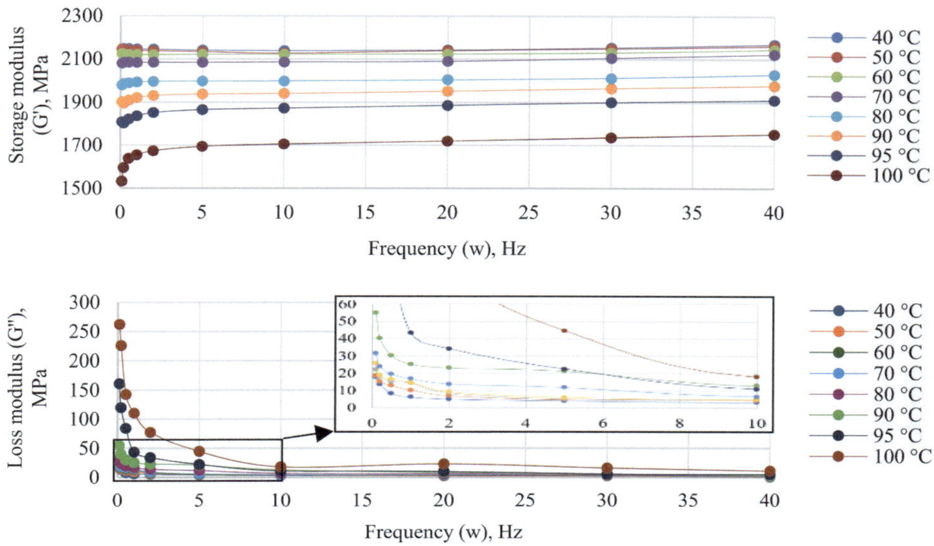

Fig. 2.13 Experimental values of the elasticity and viscosity of PP plastic

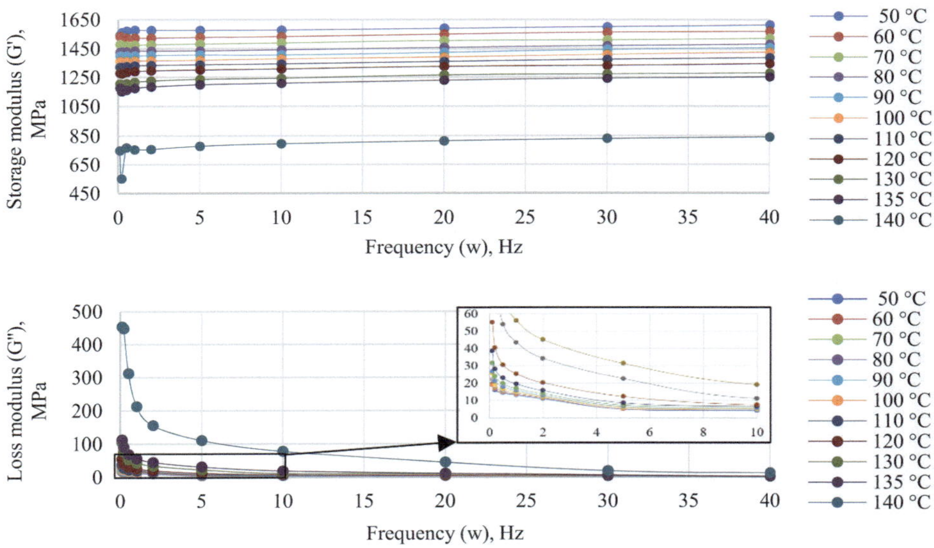

Fig. 2.14 Experimental values of the elasticity and viscosity of PETG plastic

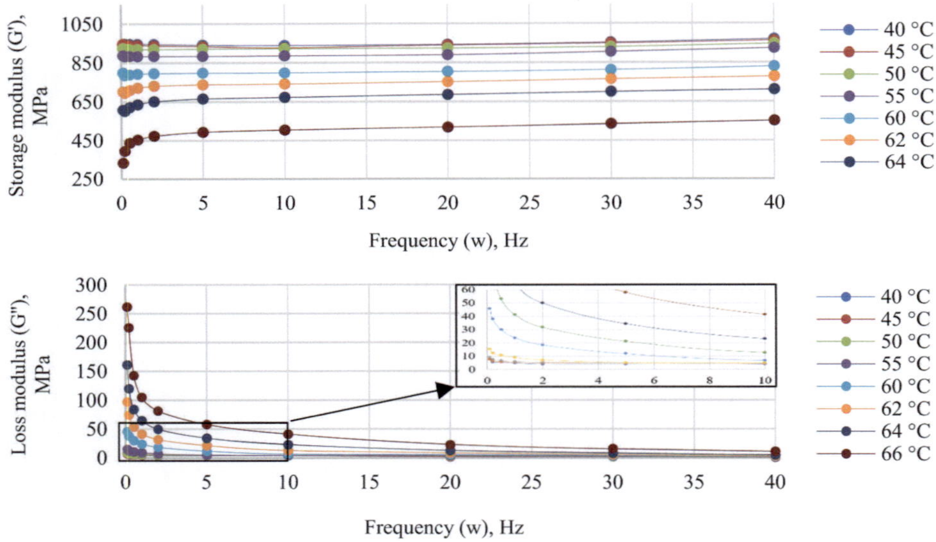

Fig. 2.15 Experimental values of the elasticity and viscosity of PVC plastic

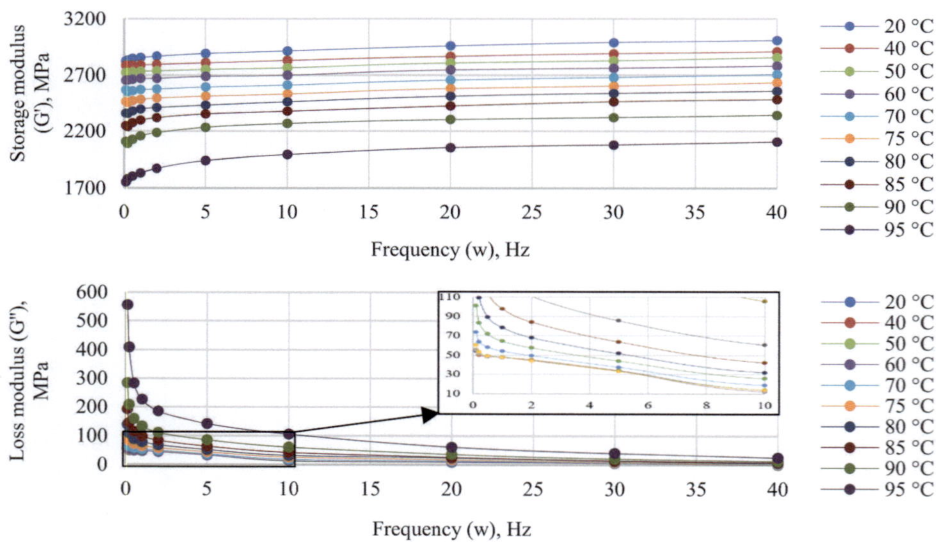

Fig. 2.16 Experimental values of the elasticity and viscosity of SAN plastic

is going to be applied during the hot stamping forming process, the material's behaviour values from the experiment needed to be increased to the needed frequency. For this, the temperature-frequency superposition (TTS) method was used. Through this study, the boundaries of unknown temperature frequencies could be expanded and the rheological dependence of the polymer on the behaviour of the material could be determined [44].

To analyse the curve obtained by the principle of superposition at high frequencies, the curve had to be extrapolated using a fitting function with selected coefficients. A sigmoid function was used to extrapolate this type of curve to which coefficients were applied. The master curves of elasticity and viscosity had to be approximated with the Kramers–Kronig relation using a sigmoid function with coefficients [45]. Sigmoid functions are presented below:

$$G'(w) = a \cdot \tan h(b(\log(w) + c)) + d \quad (2.3)$$

$$G''(w) = \frac{\pi ab}{2} \cdot \sec h\big(b(\log(w) + c)^2\big) \quad (2.4)$$

where *a*, *b*, *c*, *d*—fit coefficient; log(*w*)—natural logarithm; tan*h*—hyperbolic tangent.

After extrapolation, fit function coefficients with goodness of fit were obtained. These fit coefficients were calculated using two different programmes: *Excel* (*Microsoft*, JAV) and *MCalibration* (*PolymerFEM*, JAV). The obtained values were compared with each other by choosing the optimal ones. The values obtained are presented in Table 2.3.

A basic elasticity–viscosity curve was determined over a wide variety of frequencies after the coefficients were obtained and the values were entered into the equation. The curves are shown in Figs. 2.17, 2.18, 2.19 and 2.20.

To gain an understanding of the behaviour and performance of materials under specific conditions, this research evaluated the viscoelastic characteristics of polymers at various temperatures and dynamic loads. The storage and loss modulus, among other parameters required for further simulations, could potentially be found using the experimental DMA method. The findings of the study and the approximation function coefficients contribute to the understanding of how polymers behave under different conditions. For scientists and engineers who need to learn more about and apply these plastic materials in a variety of industrial applications, all this information could be valuable.

2.4 Vibration-Based Modelling of the Hot Stamping Process

During the ultrasonic hot stamping process, high-frequency longitudinal waves are used in addition to the embossing load. When it propagates through a material, the ultrasonic vibrational energy is absorbed and converted into another type of energy. Since thermoplastic is an amorphous material, most of the energy of ultrasonic vibrations is converted into heat. The amount of heat released increases the temperature of the mold and the

Table 2.3 Fit function coefficients with the goodness-of-fit

Temperature, °C	a	b	c	d	R^2
PP thermoplastic					
40	1017.001	0.4952	3.356	1190.173	0.976
70	1024.202	0.5166	3.362	1115.141	0.979
100	874.169	0.4232	2.988	922.355	0.985
PETG thermoplastic					
50	778.948	0.3603	3.762	881.849	0.989
100	684.086	0.3978	3.733	739.915	0.986
140	442.876	0.2332	3.146	464.954	0.980
PVC thermoplastic					
40	446.385	0.3721	4.522	533.350	0.996
55	432.533	0.3270	4.152	499.742	0.996
66	305.997	0.3780	1.315	304.552	0.991
SAN thermoplastic					
40	1400.215	0.2484	4.496	1740.810	0.980
60	1309.777	0.3918	3.504	1511.849	0.990
95	918.824	0.5472	2.192	1146.626	0.980

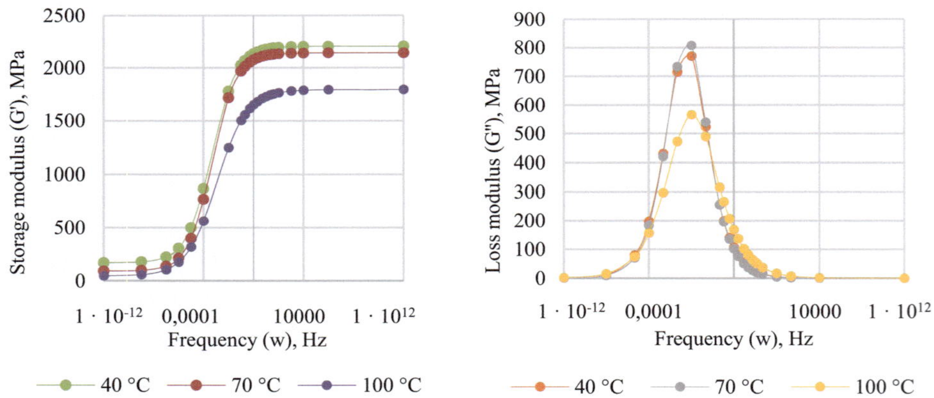

Fig. 2.17 Basic theoretical curves of PP plastic

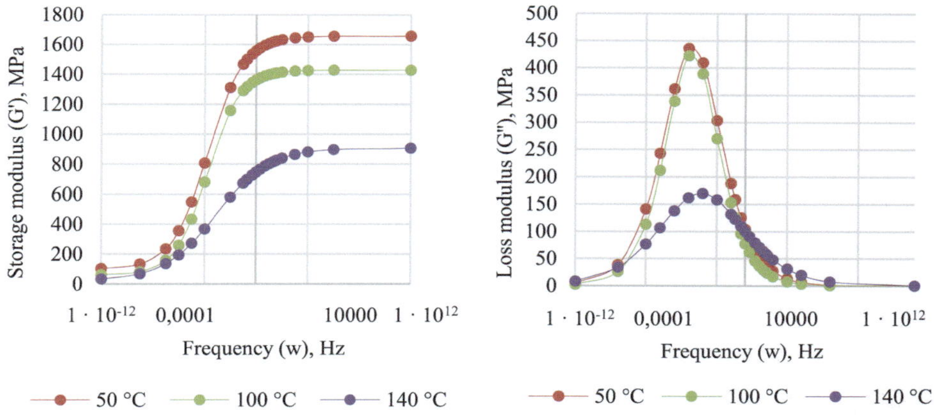

Fig. 2.18 Basic theoretical curves of PETG plastic

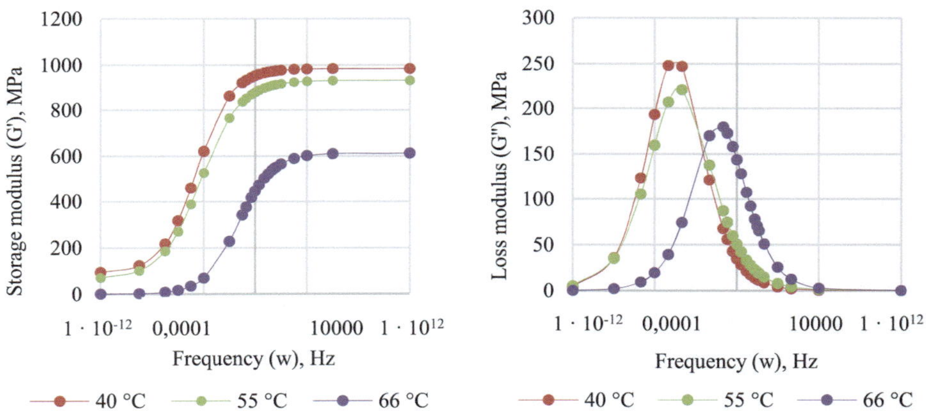

Fig. 2.19 Basic theoretical curves of PVC plastic

thermoplastic, which improves the flow of the microstructure into the mold and allows to reduce the molding force. To model and analyse thermoplastic forming using ultrasonic vibrations during hot stamping, it is necessary to determine the amount of heat that is absorbed when the ultrasonic energy is converted to thermal energy.

In Sect. 2.3, the loss modulus G'' was calculated, which represents the energy dissipated by the material. If the dissipated energy is converted into heat, then the rate of heat generation is equal to [46]:

$$\dot{q} = w \cdot \pi \cdot G'' \cdot \epsilon_0^2 \tag{2.5}$$

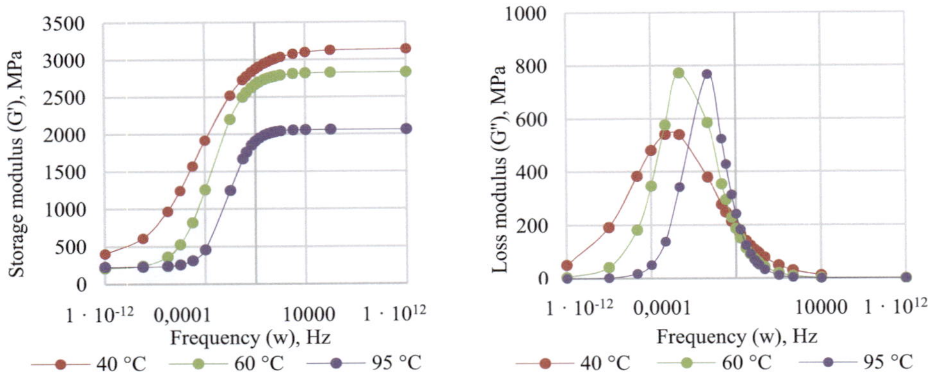

Fig. 2.20 Basic theoretical curves of SAN plastic

where w—frequency, Hz; G''—loss modulus, Pa; ϵ_0—vibration amplitude, mm.

To evaluate the difference between the traditional hot stamping method and the method with vibration, it is necessary to supplement the forming method with the inclusion of ultrasonic vibrations. For this purpose, an additional Maxwell model is available, incorporating more Maxwell elements.

Finite elements were used to analyse the technology for formation microstructures by ultrasonic hot stamping. The resulting coefficients of elasticity and viscosity had to be converted into Prony coefficients [47].

$$G'(w) = E_0\left[1 - \sum_{i=1}^{N} g_i\right] + E_0 \sum_{i=1}^{N} \frac{g_i \cdot T_i^2 \cdot w^2}{1 + T_i^2 \cdot w^2} \tag{2.6}$$

$$G''(w) = E_0 \sum_{i=1}^{N} \frac{g_i \cdot T_i^2 \cdot w^2}{1 + T_i^2 \cdot w^2} \tag{2.7}$$

Following the use of the provided loss and storage modules, the Prony series coefficients were obtained. To determine the values, the N values and the prediction values for T_i, g_i, and E_0 were selected. The G' and G'' values were calculated and then compared with the experimental results. Furthermore, the process was repeated when the R^2 values were computed and the R^2 value approached the nearest value of 1. Next, in order to obtain the value that was closest to the experimental curve, 8 Prony series terms were selected. Prony values were obtained using the *MCalibration* programme (*PolymerFEM*, JAV) and used in the finite element model. On the basis of the obtained results, a finite element model was created using *Ansys* software (*Ansys*®, Pennsylvania, United States). A viscoelastic material model was selected. The material model was divided into 1704 finite elements. In the analysis, the mold moved 4 μm into the plastic in 8 s, then was held for 1 s, and retracted in 1 s. The simulation results are presented in Fig. 2.21.

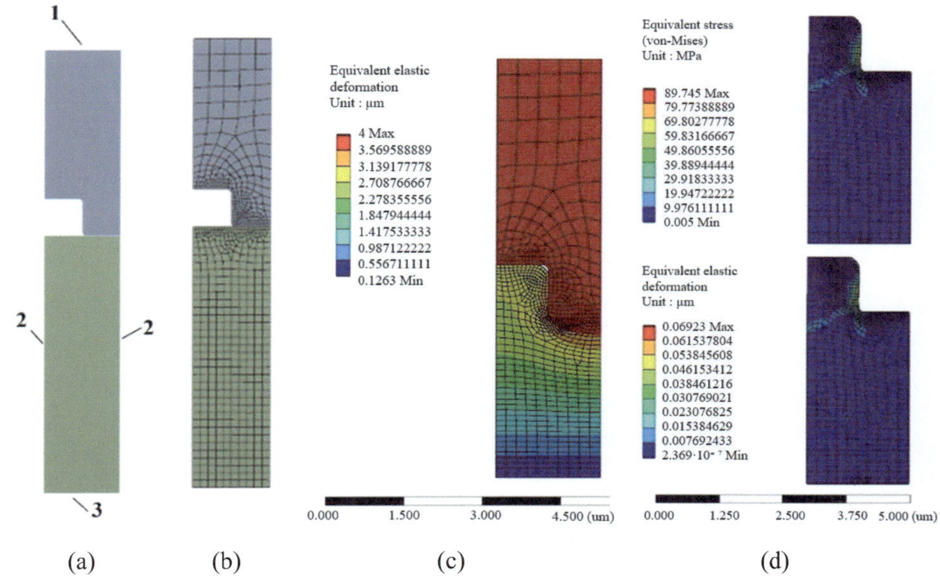

Fig. 2.21 Analysis of finite element: **a** model with boundary conditions: 1—displacement, 2—frictionless support, 3—fixed support; **b** meshed model; **c** image of deformation; **d** images of equivalent stress and elastic strain

To verify the suitability of the grid, the grid-independent verification study was carried out. The stress and strain result no longer changed with the medium density mesh. The further densification of the grids had no significant effect on the results, so the medium density grid was suitable for further studies. The results of the reliability test are presented in Table 2.4.

Throughout the simulation, the average stresses of the plastic model were presented in stress diagrams for various polymers. As the temperature increased, the relative stresses began to decrease. Although the deformation remained unchanged, it was observed that the stresses began to decrease in the eighth second. Because the stress graph was more uniform, the graphs also demonstrated how much more plastically the material behaves at higher temperatures. The findings are presented in Figs. 2.22, 2.23, 2.24 and 2.25.

Table 2.4 Grid-independent verification

	Elements	Strain, μm/μm	Stress, MPa
Coarse	791	0.12095	701.15
Medium	1704	0.12412	421.78
Fine	6066	0.12129	431.74

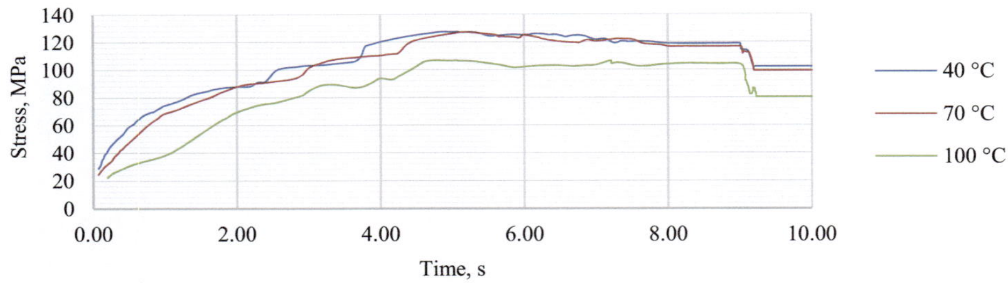

Fig. 2.22 Results of PP thermoplastic modelling

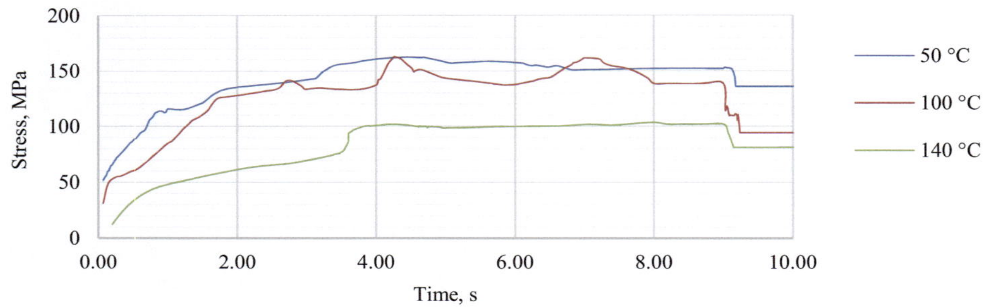

Fig. 2.23 Results of PETG thermoplastic modelling

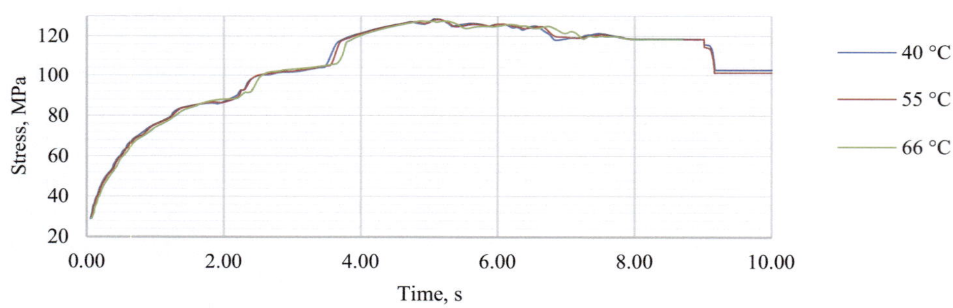

Fig. 2.24 Results of PVC thermoplastic modelling

In conclusion, a promising technology in the plastic molding process can be stated to be the ultrasonic hot stamping process. The effective transformation of the ultrasonic vibrational energy into thermal energy made possible by this technique improves the flow of plastic into the mold. The behaviour of this process has been better understood, and key factors that characterize this behaviour have been identified, because of experimental

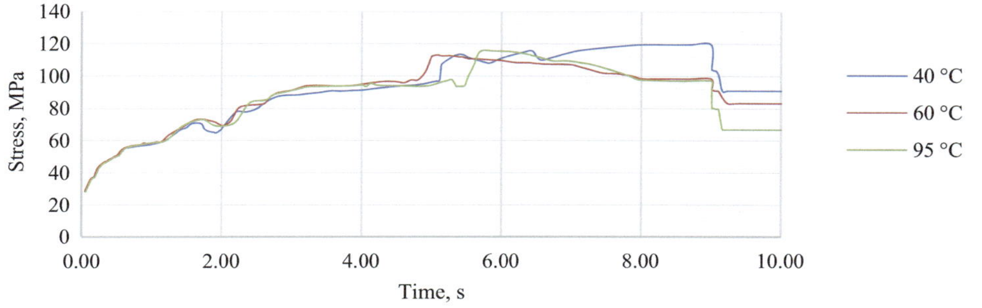

Fig. 2.25 Results of SAN thermoplastic modelling

research and finite element modelling. Future research aimed at improving the efficiency of plastic molding and optimizing the ultrasonic hot stamping process could greatly benefit from these models. It is essential to consider the many aspects that impact the process, such as material, ultrasonic vibration settings, and geometry, since they could have an impact on the final result.

References

1. Nawrot W, Malecha K (2020) Additive manufacturing revolution in ceramic microsystems. Microelectron Int 37(2):79–85. https://doi.org/10.1108/MI-11-2019-0073
2. Niculescu AG, Chircov CB, Birca AC, Grumezescu AM (2021) Fabrication and applications of microfluidic devices: a review. Int J Mol Sci 22:2011. https://doi.org/10.3390/ijms22042011
3. Calero M, Fernández R, García P, García JV, García M, Gamero-Sandemetrio E, Reviakine I, Arnau A, Jiménez Y (2020) A multichannel microfluidic sensing cartridge for bioanalytical applications of monolithic quartz crystal microbalance. Biosensors 10:189. https://doi.org/10.3390/bios10120189
4. Dastider SG, Abdullah A, Jasim I, Yuksek NS, Dweik M, Almasri M (2018) Low concentration *E. coli* O157:H7 bacteria sensing using microfluidic MEMS biosensor. Rev Sci Instrum 89:125009. https://doi.org/10.1063/1.5043424
5. Cairone F, Davi S, Stella G, Guarino F, Recca G, Cicala G, Bucolo M (2020) 3D-Printed micro-optofluidic device for chemical fluids and cells detection. Biomed Microdevice 22:37. https://doi.org/10.1007/s10544-020-00487-3
6. Guo H, Zhao H, Niu H, Ren Y, Fang H, Fang X, Lv R, Maqbool M, Bai S (2021) Highly thermally conductive 3D printed graphene filled polymer composites for scalable thermal management applications. ACS Nano 15:6917–6928. https://doi.org/10.1021/acsnano.0c10768
7. Agrawal A, Kushwaha HM, Jadhav RS (2020) Microscale flow and heat transfer: mathematical modelling and flow physics. Springer, Cham. ISBN 978-3-030-10661-4. https://doi.org/10.1007/978-3-030-10662-1
8. Ikumapayi OM, Akinlabi ET, Adeoye AOM, Fatoba SO (2021) Microfabrication and nanotechnology in manufacturing system—an overview. Mater Today Proc 44:1154–1162. https://doi.org/10.1016/j.matpr.2020.11.233

9. Juang YJ, Chiu YJ (2022) Fabrication of polymer microfluidics: an overview. Polymers 14:2028. https://doi.org/10.3390/polym14102028
10. Melentiev R, Lubineau G (2022) Large-scale hot embossing of 1 μm high-aspect-ratio textures on ABS polymer. CIRP J Manuf Sci Technol 38:340–349. https://doi.org/10.1016/j.cirpj.2022.05.011
11. Scott SM, Ali Z (2021) Fabrication methods for microfluidic devices: an overview. Micromachines 12:319. https://doi.org/10.3390/mi12030319
12. Deshmukh SS, Goswami A (2020) Hot embossing of polymers—a review. Mater Today Proc 26:405–414. https://doi.org/10.1016/j.matpr.2019.12.067
13. Ran J, Wang X, Liu Y, Yin S, Li S, Zhang L (2023) Micro reactor-based micro/nanomaterials: fabrication, advances, and outlook. Mater Horiz 10:2343–2372. https://doi.org/10.1039/D3MH00329A
14. Ma Y, Zhang G, Cao S, Huo Z, Han J, Ma S, Huang Z (2023) A Review of advances in fabrication methods and assistive technologies of micro-structured surfaces. Processes 11(5):1337. https://doi.org/10.3390/pr11051337
15. Meng Z, Qiu Z, Shi Y, Wang S, Zhang G, Pi Y, Pang H (2023) Micro/nano metal-organic frameworks meet energy chemistry: a review of materials synthesis and applications. eScience 3(2):100092. https://doi.org/10.1016/j.esci.2023.100092
16. Moura BC, Rosero-Romo JJ, Monteiro H, Alberto AR, Laranjeira J, Martin-Iglesias J, Silvan U, Lanceros-Mendez S, Salazar D, Martins CF (2024) Addressing safety and sustainability issues in the development of nano-enabled MULTI-FUNctional materials for metal additive manufacturing. Sustain Mater Technol 41:e01085. https://doi.org/10.1016/j.susmat.2024.e01085
17. Cigane U, Palevicius A, Janusas G (2021) Review of nanomembranes: materials, fabrications and applications in tissue engineering (bone and skin) and drug delivery systems. J Mater Sci 56:13479–13498. https://doi.org/10.1007/s10853-021-06164-x
18. Sebastian EM, Jain SK, Purohit R, Dhakad SK, Rana RS (2020) Nanolithography and its current advancements. Mater Today Proc 26:2351–2356. https://doi.org/10.1016/j.matpr.2020.02.505
19. Haque SR (2023) Preparation, characterization, applications and future challenges of nanomembrane—a review. Hybrid Adv 3:100027. https://doi.org/10.1016/j.hybadv.2023.100027
20. Dong Z, He Q, Shen D, Gong Z, Zhang D, Zhang W, Ono T, Jiang Y (2023) Microfabrication of functional polyimide films and microstructures for flexible MEMS applications. Microsyst Nanoeng 9:31. https://doi.org/10.1038/s41378-023-00503-5
21. Lin Q (2023) Polymeric electronic materials for microelectronics manufacturing: a review. Polymers 286:126395. https://doi.org/10.1016/j.polymer.2023.126395
22. Du L, Zhai K, Li X, Liu S, Tao Y (2020) Ultrasonic vibration used for improving interfacial adhesion strength between metal substrate and high-aspect-ratio thick SU-8 photoresist mould. Ultrasonics 103:106100. https://doi.org/10.1016/j.ultras.2020.106100
23. Tahir U, Shim YB, Kamran MA, Kim DI, Jeong MY (2021) Nanofabrication techniques: challenges and future prospects. J Nanosci Nanotechnol 21:4981–5013. https://doi.org/10.1166/jnn.2021.19327
24. Yang L, Wei J, Ma Z, Song P, Ma J, Zhao Y, Huang Z, Zhang M, Yang F, Wang X (2019) The fabrication of micro/nano structures by laser machining. Nanomaterials 9:1789. https://doi.org/10.3390/nano9121789
25. Gao S, Huang H (2017) Recent advances in micro- and nano-machining technologies. Front Mech Eng 12:18–32. https://doi.org/10.1007/s11465-017-0410-9
26. Melentiev R, Fang F (2020) Fabrication of micro-channels on Co–Cr–Mo joints by micro-abrasive jet direct writing. J Manuf Process 56:667–677. https://doi.org/10.1016/j.jmapro.2020.05.022

27. Praveena BA, Lokesh N, Abdulrajak B, Santhosh N, Praveena BL, Vingesh R (2022) A comprehensive review of emerging additive manufacturing (3D printing technology): methods, materials, applications, challenges, trends and future potential. Mater Today Proc 52:1309–1313. https://doi.org/10.1016/j.matpr.2021.11.059
28. Camarena-Chávez VA, Castro-Beltrán R, Medina-Cázares OM, Álvarez-Martínez JU, Ramos-Ortíz G, Gutiérrez-Juárez G (2020) Implementation and assessment of a low-cost 3D laser platform controlled by open software for printing polymeric micro-structures. J Micromech Microeng 30:035010. https://doi.org/10.1088/1361-6439/ab6c75
29. Jaksic Z, Jaksic O (2020) Biomimetic nanomembranes: an overview. Biomimetics 5:24. https://doi.org/10.3390/biomimetics5020024
30. Xia DP, Lai LJ (2019) Fabrication of a microstructure array using electrochemical wet stamping technique with a polyacrylamide gel. Int J Electrochem Sci 14:3434–3442. https://doi.org/10.20964/2019.04.20
31. Liu H, Li Y, Lin W, Hong M. High-aspect-ratio crack free microstructures fabrication on sapphire by femtosecond laser ablation. Opt Laser Technol 132:106472. https://doi.org/10.1016/j.optlastec.2020.106472
32. Wu K, He X, Wang J, Pan T, He R, King F, Cao Z, Ju F, Huang, Z, Nie L (2022) Recent progress of microfluidic chips in immunoassay. Front Bioeng Biotechnol 10. Prieiga per: https://doi.org/10.3389/fbioe.2022.1112327
33. Zhao B, Qiang Y, Wu W, Jiang B (2021) Tuning power ultrasound for enhanced performance of thermoplastic micro-injection molding: principles, methods, and performances. Polymers 13:2877. https://doi.org/10.3390/polym13172877
34. Liang F, Qiao Y, Duan M, Ju A, Lu N, Li J, Tu J, Lu Z (2018) Fabrication of a microfluidic chip based on the pure polypropylene material. RSC Adv 8:8732–8738. https://doi.org/10.1039/C7RA13334K
35. Mehta V, Sudhakaran SV, Rath SN (2021) Facile route for 3D printing of transparent PETg-based hybrid biomicrofluidic devices promoting cell adhesion. ACS Biomater Sci Eng 7:3947–3963. https://doi.org/10.1021/acsbiomaterials.1c00633
36. Li Y, Li K, Gong F (2021) Fabrication and optical characterization of polymeric aspherical microlens array using hot embossing technology. Appl Sci 11:882. https://doi.org/10.3390/app11020882
37. Cui F, Jafarishad H, Zhou Z, Chen J, Shao J, Wen Q, Liu Y, Zhou HS (2020) Batch fabrication of electrochemical sensors on a glycol-modified polyethylene terephthalate-based microfluidic device. Biosens Bioelectron 167:112521. https://doi.org/10.1016/j.bios.2020.112521
38. Kojic SP, Stojanovic GM, Radonic V (2019) Novel cost-effective microfluidic chip based on hybrid fabrication and its comprehensive characterization. Sensors 19:1719. https://doi.org/10.3390/s19071719
39. Voicu D, Lestari G, Wang Y, Debono M, Seo M, Cho S, Kumachevaac E (2017) Thermoplastic microfluidic devices for targeted chemical and biological applications. RSC Adv 7:2884–2889. https://doi.org/10.1039/C6RA27592C
40. Ashok M, Deepika S, Sowndharya P, Muthukumar K (2019) Cotton candy driven chitosan and gelatin coated poly(styrene-co-acrylonitrile) microfibers for anti-microbial wound dressing applications. Mater Res Express 6:125339. https://doi.org/10.1088/2053-1591/ab5b2e
41. Zhang Y, Zhang X, Cao Y, Feng J, Yang W (2021) Acrylonitrile-styrene-acrylate particles with different microstructure for improving the toughness of poly(styrene-co-acrylonitrile) resin. Adv Polym Technol 2021:3004824. https://doi.org/10.1155/2021/3004824

42. Shirazi MMA, Bazgir S, Meshkani F (2020) A dual-layer, nanofibrous styrene-acrylonitrile membrane with hydrophobic/hydrophilic composite structure for treating the hot dyeing effluent by direct contact membrane distillation. Chem Eng Res Des 164:125–146. https://doi.org/10.1016/j.cherd.2020.09.030
43. Ciganas J, Griskevicius P, Palevicius A, Urbaite S, Janusas G (2022) Development of finite element models of PP, PETG, PVC and SAN polymers for thermal imprint prediction of high-aspect-ratio microfluidics. Micromachines 13:1655. https://doi.org/10.3390/s151229876
44. Behera A (2021) Magnetostrictive materials. In: Advanced materials. Springer, Cham, pp 127–156. https://doi.org/10.1007/978-3-030-80359-9_4
45. Zeltmann SE, Prakash KA, Doddamani M, Gupta N (2017) Prediction of modulus at various strain rates from dynamic mechanical analysis data for polymer matrix composites. Compos Part B Eng 120:27–34. https://doi.org/10.1016/j.compositesb.2017.03.062
46. Wang W, Zeng Y (2020) Polypropylene—polymerization and characterization of mechanical and thermal properties. IntechOpen. ISBN 978-1-83880-416-9. https://doi.org/10.5772/intechopen.73995
47. Li L, Li W, Wang H, Zhao J, Wang Z, Dong M, Han D (2018) Investigation of Prony series model related asphalt mixture properties under different confining pressures. Constr Build Mater 166:147–157. https://doi.org/10.1016/j.conbuildmat.2018.01.120

Development of a Magnetostrictive Transducer Used for Microstructure Formation

3

3.1 Theoretical Analysis of the Forming Tool

The magnetostriction process is a physical phenomenon in which a ferromagnetic material is deformed under the influence of an external magnetic field [1]. Due to the ability to control the magnetic field, it is possible to extract the vibration process in a wide frequency range. Typically, a magnetostrictive transducer generates a large amount of heat during vibration, which is often unwanted and is eliminated through design solutions and cooling agents. In the case of the technology under development, the released heat is used in the hot stamping process.

Similar studies were carried out using piezoelectric ultrasonic devices. Researchers have found that ultrasonic vibrations improve the formation process. During formation, the contact formation time and the required formation force were found to be reduced [2]. The forming tool being developed in this research will be different because it will not have an additional heating element, and the transducer will be magnetostrictive rather than piezoelectric. Mathematical models and equations for heat exchange and magnetostrictor deformations are given below. *COMSOL Multiphysics 5.4* software (*COMSOL*®, Burlington, United States) was used for the analysis. The mathematical model of the magnetostrictive device is presented in Fig. 3.1.

Induction heating is a process that does not require contact to transfer energy and heat the body. This process uses high-frequency electricity, and the induction process can only heat electrically conductive material. This heating process is very efficient because it heats the material from the inside [3].

Since heat exchange occurs in the system, a heat exchange formula equivalent to induction heating is used to define the system. The heat transfer formula is a parabolic partial differential equation that describes the temperature distribution of a certain area at certain moments of time. The formula for heat exchange is given below:

A. Palevicius et al., *Nano/Micro Functional Elements Formation for Bioengineering Applications*, Synthesis Lectures on Biomedical Engineering,
https://doi.org/10.1007/978-3-031-81509-6_3

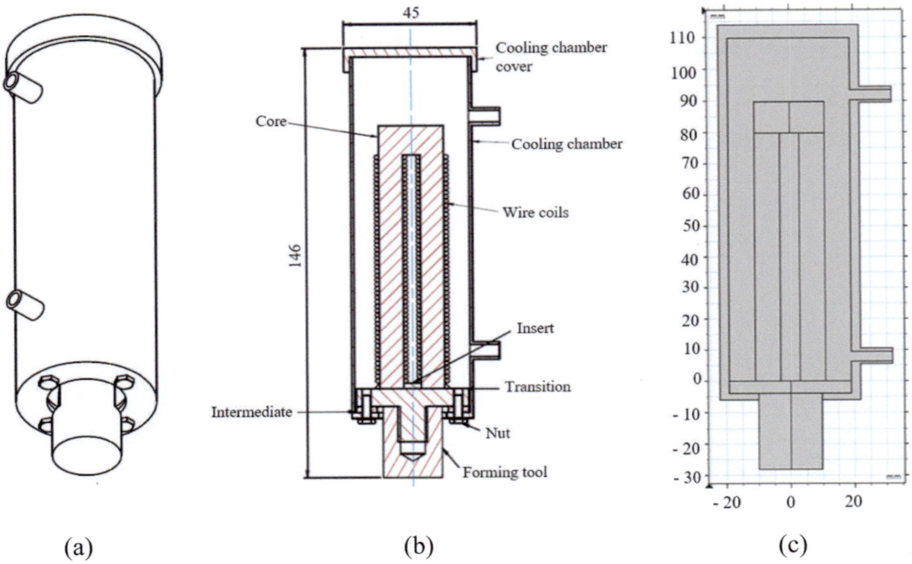

Fig. 3.1 Magnetostrictive transducer: **a** isometric view; **b** cross-sectional view; **c** 2D representation of a mathematically simplified device

$$\rho C_p\left(\frac{\partial T}{\partial t}\right) - \nabla k \nabla T = Q(T) \tag{3.1}$$

where ρ—density, kg/m^3; C_p—specific heat capacity, J/(kg K); k—thermal conductivity, W/(m K); Q—heat source density function.

The material parameters are not linear equations with temperature, so the variables were defined over a range of temperatures. Parameter dependencies were plotted using non-linear equations. The parameters of the materials are presented in Fig. 3.2.

Using equations and the parameter values given in Fig. 3.2, it is possible to determine the temperature variation of the model depending on the ongoing process. *COMSOL Multiphysics 5.4* two-dimensional space (*COMSOL*®, Burlington, United States) was chosen to model the process. The model was divided into finite triangular elements. A maximum element size of 2.84 mm and a minimum of 0.0107 mm were established. Room temperature (293.15 K) was chosen as the starting temperature for the analysis. The heat source was selected to be the core, and the air was used for heat exchange, with a heat transfer coefficient of 25.32 W/(m^{2}K). The specific heat ratio of steel was 1.

First, the forming tool was designed with a cooling chamber. A heat transfer analysis showed that a magnetic flux density of 1 T was required to achieve a temperature of 210 °C. Creating such a magnetic flux required a powerful power supply, so it was decided that a cooling chamber would not be used. After an improved heat exchange analysis, the

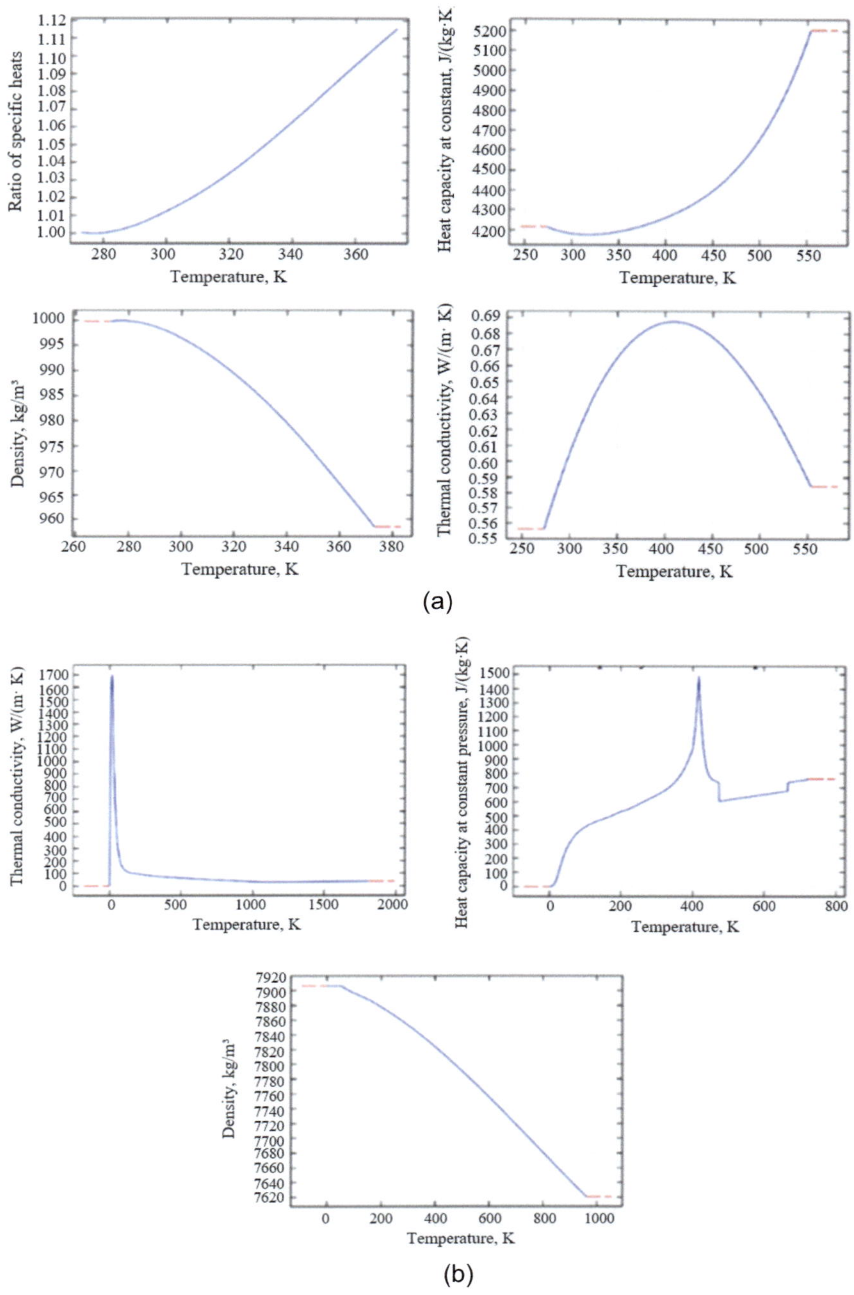

Fig. 3.2 Dependences of properties on temperature: **a** water; **b** steel [4]

results showed that a temperature of 225 °C could be generated at a magnetic flux density of 0.6 T. Even higher temperatures could be achieved with a stronger magnetic field, but since plastics were used in the molding process and their melting point often did not exceed 200 °C, a molding temperature of 225 °C was sufficient for this study. It was also possible to see that heat exchange between metals was much more intense than between air. Different characteristics of the materials made it possible to reach a state where the gap between the magnetostrictor and the forming tool heats up most intensively. The aluminum forming tool ensured temperature uniformity between the forming tool and the intermediate element. The results are presented in Fig. 3.3.

One coil in the magnetostrictor was designed to create the magnetic field and the other to generate the magnetostrictive process. The goal of modelling the magnetostrictive process was to determine what deformations could be obtained using this process. To find the deformations, it was necessary first to find the stresses created by the magnetostrictive process. The stresses in this process were calculated using the formula:

$$S = C_H[\varepsilon_{el} - \varepsilon_{me}(M)] \tag{3.2}$$

where C_H—stiffness matrix; ε—deformation, mm/mm; $\varepsilon_{\mathrm{me}}$—magnetostrictive deformation, mm/mm.

The magnetostrictive strain was calculated using a quadratic isotropic function of the magnetic field:

$$\varepsilon_{me} = \left(3\lambda_s/2M_s^2\right)dev\left(M_iM_j\right) \tag{3.3}$$

where λ_s—magnetostrictor saturation value, (m/m)/(A/m); M_s—saturation magnetization, A/m.

The magnetostrictive permendur 49 ($Co_{49}Fe_{49}V_2$) alloy was used in the process. The properties of this magnetostrictive material are presented in Table 3.1.

The simulation was performed using the given formulas and parameter values. The mathematical model composed of finite elements. The current density was chosen from zero to 1×10^8 A/m^2. The core, transition, and forming tools were selected as the domain of study, since the magnetostrictive process occurred in the core. This type of analysis was called a Multiphysics task because it included mechanical deformation and magnetic field analysis. After the simulation, the deformation image and the non-linear curve were obtained, which is presented in Fig. 3.4.

After analysing the results, it was found that even with the magnetostriction coils located to one side, the deformations throughout the structure are uniform and symmetric because of the uniform distribution of the magnetic field. It is shown in the graph of the magnetostrictive deformation's dependence on current density that the magnetostrictor has non-linear characteristics. According to the overview of the results, the recently developed

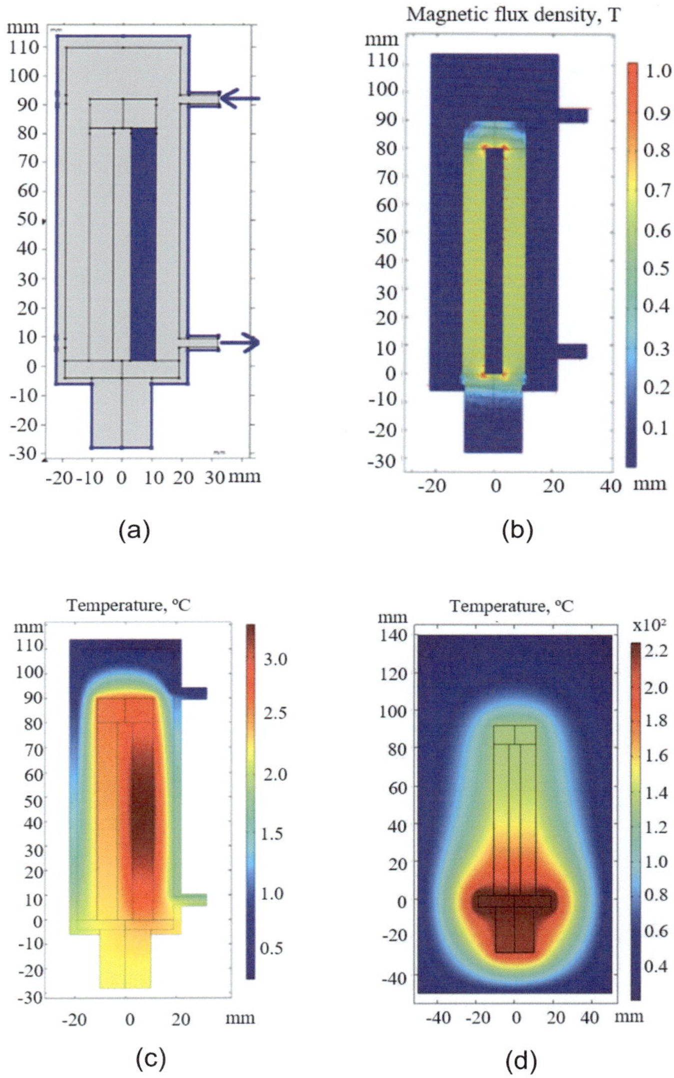

Fig. 3.3 Heat exchange analysis: **a** mathematical model; **b** magnetic flux density; **c** temperature distribution with cooling system; **d** temperature distribution without cooling

microstructure formation tool was able to achieve deformations of up to 4.5 μm and a formation temperature of up to 200 °C. Modifying the generated current's frequency may control the strain's frequency, while varying the alternating current's magnetization or intensity can affect the strains' amplitude.

Table 3.1 Properties of magnetostrictive material [4]

Properties	Value	Unit
Young's modulus	60×10^9	Pa
Poisson's ratio	0.45	–
Density	7870	kg/m^3
Electrical conductivity	5.96×10^6	s/m
Thermal resistance	45	W/(m K)
Specific heat capacity	510	J/(kg K)
Magnetic saturation	1.5×10^6	A/m
Magnetic susceptibility	200	–
Magnetostrictive saturation	200	ppm

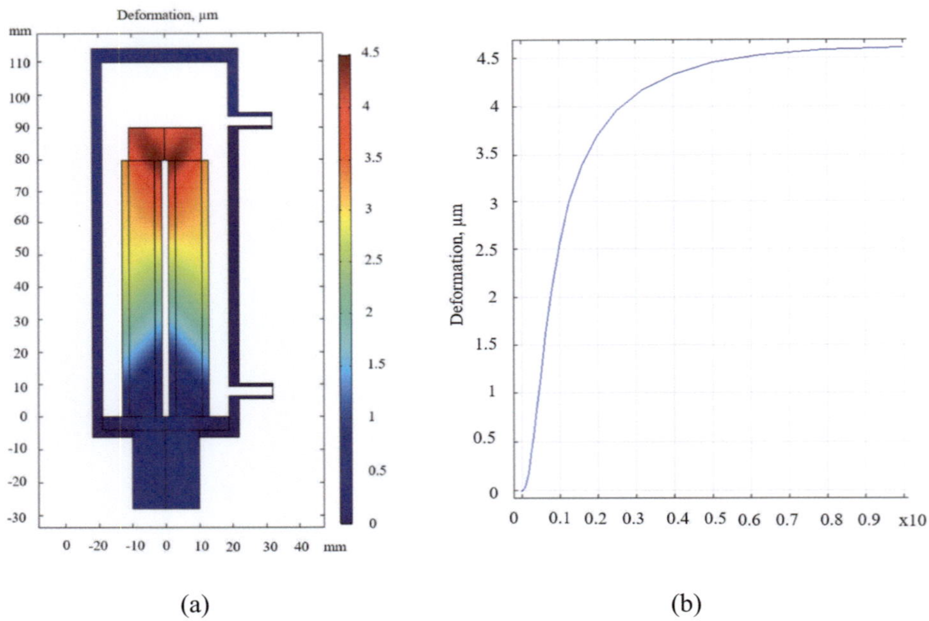

Fig. 3.4 Simulation results of magnetostrictor deformations: **a** image of magnetostrictor deformation; **b** dependence of magnetostrictive strain on current density

3.2 Development and Analysis of Magnetostrictive Transducer

Various types of technologies were used to create the different parts of the microstructuring tool. An alloy of steel known as permendur 49 was used to construct the tool's core. After being formed with a computerized laser cutting machine, the structure was

Fig. 3.5 Magnetostrictive transducer

riveted. The magnet wire, composed of a ferromagnetic material, was attached to the welded transition. Two different wires with a cross section of 0.75 mm^2 were used to surround the core. The wires were resistant to temperatures up to 180 °C. Two wires were used: one for the inductive process and the other for magnetization. Aluminum was used to produce the forming tool. The decision had been made to use the transition as the pressure point of the external force because of the design. According to this, the tube that was connected to the transition included the magnetostrictor core. The magnetization and induction process wires were inserted through a hole made on the tube's side. The image of the magnetostrictor is presented in Fig. 3.5.

The magnetostrictor was controlled by additional equipment, which also guaranteed all required criteria. An inductive process was created using an amplifier, a resistor, and a signal generator (*UNI-T UTG2025A*, *UNI-TREND*, Germany). A 10 A constant current source was employed to induce magnetization. This generator creates a sinusoidal signal that was up to 2.5 V and sends it to the amplifier. The signal was intensified using a *Wilcoxon PA8HF* amplifier (*KDP*, United Kingdom). A resistance of 4 Ω had to be added to the circuit for this amplifier to reach its maximum power output of 600 W. Since the magnetostrictor winding resistance for the inductive process could be as low as zero, an extra high-power resistance of 4 Ω was added. Fans and liquid-cooled radiators were used to ensure that the resistors were sufficiently cooled.

The developed device was tested when the whole test equipment platform was completed. The first step in evaluating the forming tool was to analyse the temperature changes when the forming parameters were changed. A study was performed to determine the effect of magnetization on the forming temperature. The findings are shown in Fig. 3.6.

The results of the analysis showed that the intensity of temperature variation was decreased by magnetization. The temperature change was examined during the magnetization-free experiment. However, the experiment had to be interrupted at the

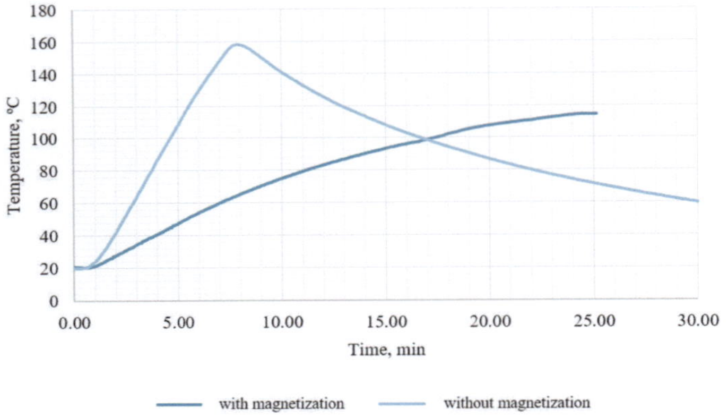

Fig. 3.6 Curves of temperature change

eighth minute to prevent damage to the windings' thermal protection because of an extreme increase in temperature.

A thermal camera had been used in the temperature research to observe variations in the device's temperature. The *FLIR T420* thermal camera (*Flir*, Oregon, USA) was selected for this purpose. The equipment of the experiment is presented in Fig. 3.7.

During data collection and analysis, it was observed that various temperature distributions were impacted by the transition caused by the ferromagnetic material. Because

Fig. 3.7 Equipment for the temperature experiment: 1—resistance, 2—DC power supply, 3—signal generator, 4—amplifier, 5—magnetostrictive transducer, 6—thermal camera

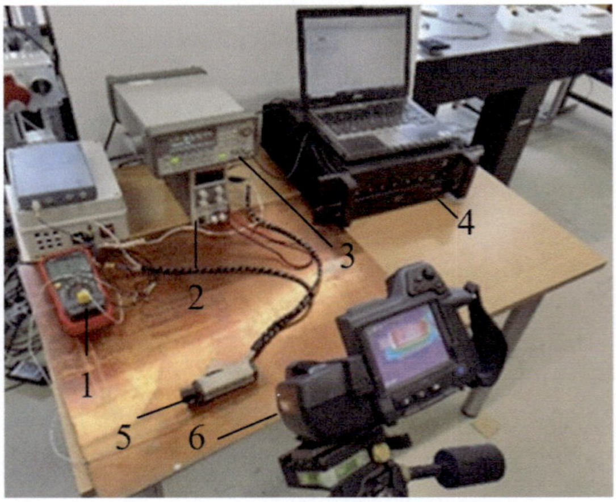

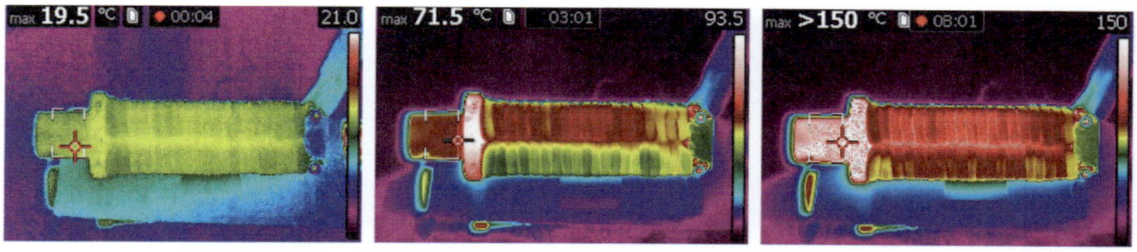

(a) (b) (c)

Fig. 3.8 Temperature distribution in magnetostrictive transducer: **a** temperature distribution at the initial moment of time; **b** temperature distribution after 3 min; **c** temperature distribution after 8 min

the thermal exchange properties of the aluminium forming tool were superior to those of the magnetostrictive material, its temperature was near the transition. Additionally, it has been determined that the magnetization varies less than that of the induction coil. The fact that there was more energy flowing during the induction process than during the magnetization phase may have had an impact on this. The temperature distribution in the magnetostrictive transducer is presented in Fig. 3.8.

The operational vibration modes were analysed using the *PRISM* holographic system (*Hytec*, Los Alamos, United States). The operating principle of the *PRISM* system is based on the principle of interferometry of the beams of two laser sources. One laser source is objective and directed at the magnetostrictor, whereas the other beam is supported and emanates directly from the video camera. The interference of these two laser lights is captured by a camera and displayed on a computer screen. The results obtained provide information on the forms of magnetostrictor oscillations. The working concept of the *PRISM* system with the tested equipment is shown in Fig. 3.9.

One end of the magnetostrictive transducer was rigidly attached with a stand to the base, allowing the forming tool to vibrate freely. Such an attachment made it possible to change the excitation frequency in real time and observe how the magnetostrictive transducer responds to changes. First, the magnetostrictive transducer was tested without a forming tool to determine additional structural vibrations. The results of the vibration experiment are presented in Fig. 3.10.

Without the forming tool, the geometry vibrates inaccurately. After adding the forming tool to the structure and changing the excitation frequency from 12.05 to 10.1 kHz, the forming tool began to oscillate in the normal direction as intended. The main drawback of this experiment is that the *PRISM* system cannot capture vibration in the tangential direction, so *Polytech* equipment (*Polytech PSV-500-3D-HV*, *Polytech*, Dieburg, Germany) was additionally used for this purpose. Furthermore, the magnetostrictive transducer and vibration modes were analysed using a *Polytech* vibrometer. The scanning laser system provided an opportunity to visualise vibration shapes without thermal influence and to observe the vibration shape in a three-dimensional space, that is, to observe tangential and normal vibrations. Vibration analysis using the *Polytech* system is presented in Fig. 3.11.

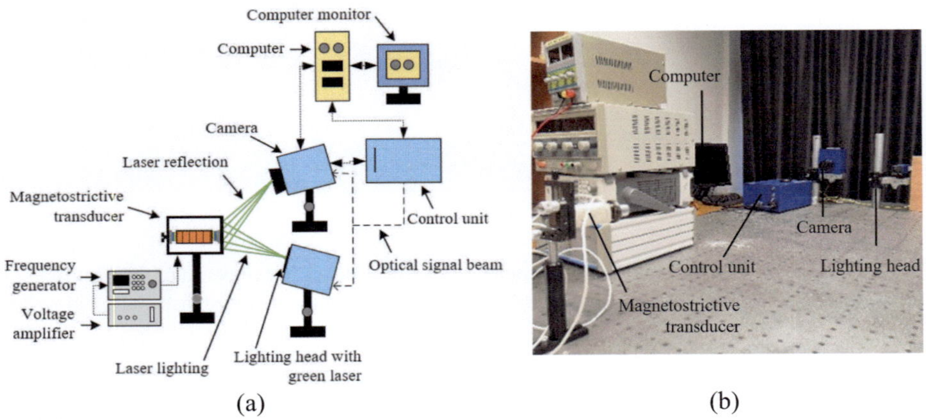

Fig. 3.9 Vibration analysis using the PRISM system: **a** schematic diagram of the PRISM system; **b** photograph of the experiment

The equipment used finite elements that allowed seeing the shapes of vibrations in space regardless of its size, and vibrations could be analysed up to a frequency of 1 GHz. Oscillation velocity amplitudes at different frequencies were obtained using this system. The results of the experiment are presented in Fig. 3.12.

An analysis of *Polytech PSV-500-3D-HV* demonstrated that the system's oscillation amplitude was highest at the resonant frequency of 10.0625 kHz. From Fig. 3.12, at lower frequencies the oscillations occurred in various tangential directions. At the resonant frequency, the amplitude of oscillations in the direction normal to the surface was the largest and in the tangential directions the minimum.

When using a three-dimensional laser doppler, only the speed of the oscillations is determined, but the amplitude of the oscillations is not indicated. *Polytech* equipment was used to determine the amplitude, which used only one laser beam. This system allowed us to convert the graph of light variation into deformations of the order of micrometres. The images and results of the experiment are presented in Fig. 3.13.

An oscillation experiment was performed during which the magnetization conditions were changed. First, the deformations were measured when the magnetization circuit was open; then, the change was measured when the circuit was closed and the magnetization voltage changed. Although the temperature of the magnetostrictor and the precise resonant frequency changes during the experiment. Therefore, the excitation frequency was continuously modified to maintain the process in the resonant mode. The collected data demonstrated that, when the magnetization intensifies, the oscillations' amplitude rises.

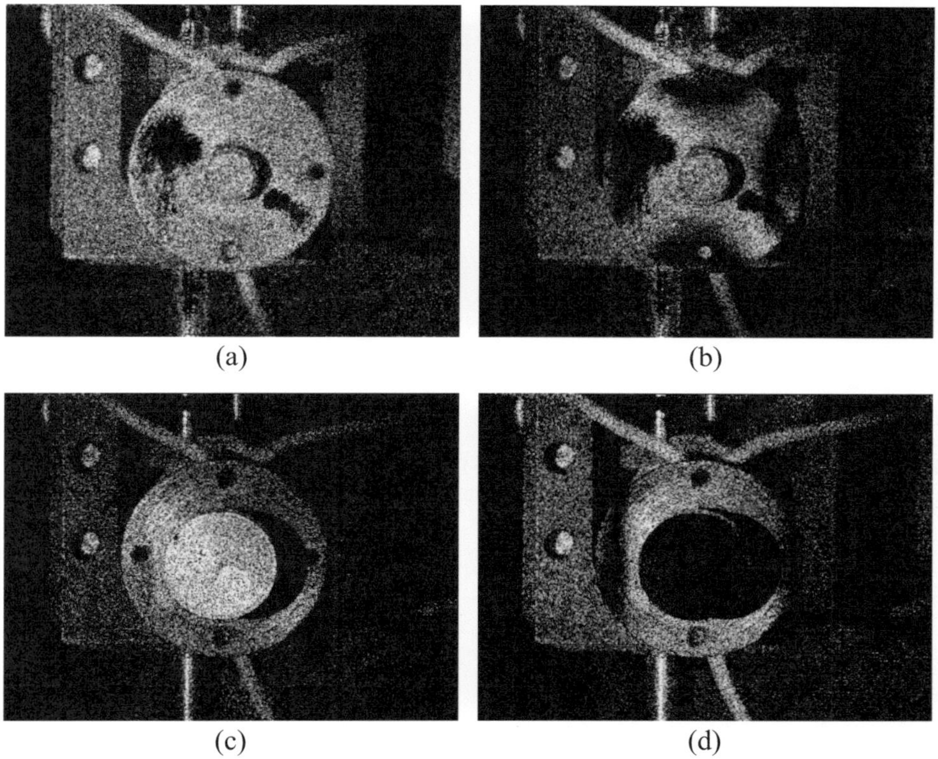

Fig. 3.10 Results of vibration analysis: **a** vibration shape without shaping tool and without excitation; **b** waveform without shaping tool at 12.05 kHz excitation frequency; **c** waveform with shaping tool and without excitation; **d** waveform with shaper at 10.1 kHz excitation frequency

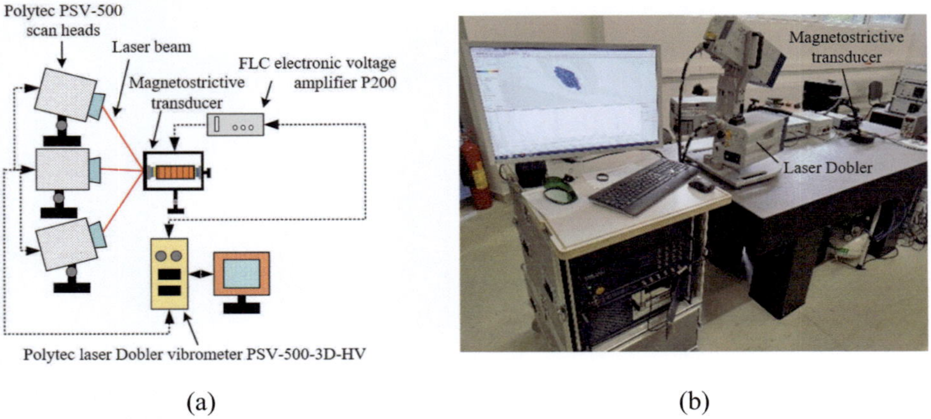

Fig. 3.11 Vibration analysis using the Polytech PSV-500-3D-HV system: **a** principle of the Polytech system; **b** experiment of magnetostrictive transducer using a Polytech laser doppler vibrometer

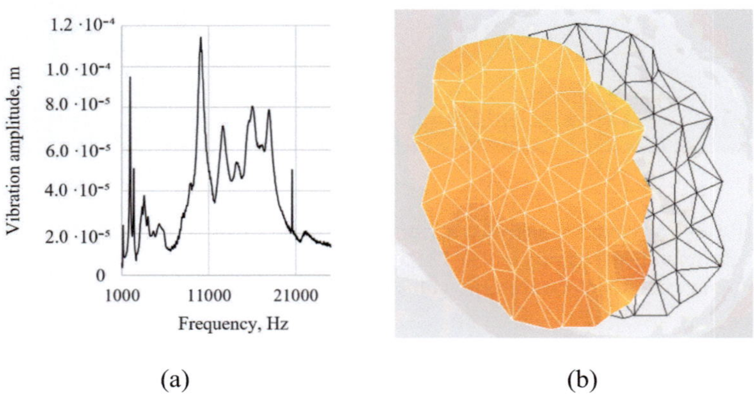

Fig. 3.12 Vibration analysis using a Polytech PSV-500-3D-HV system: **a** plot of vibration amplitudes and frequencies; **b** finite element decomposition of the shaping tool representing the amplitude of oscillation at 10.0625 kHz

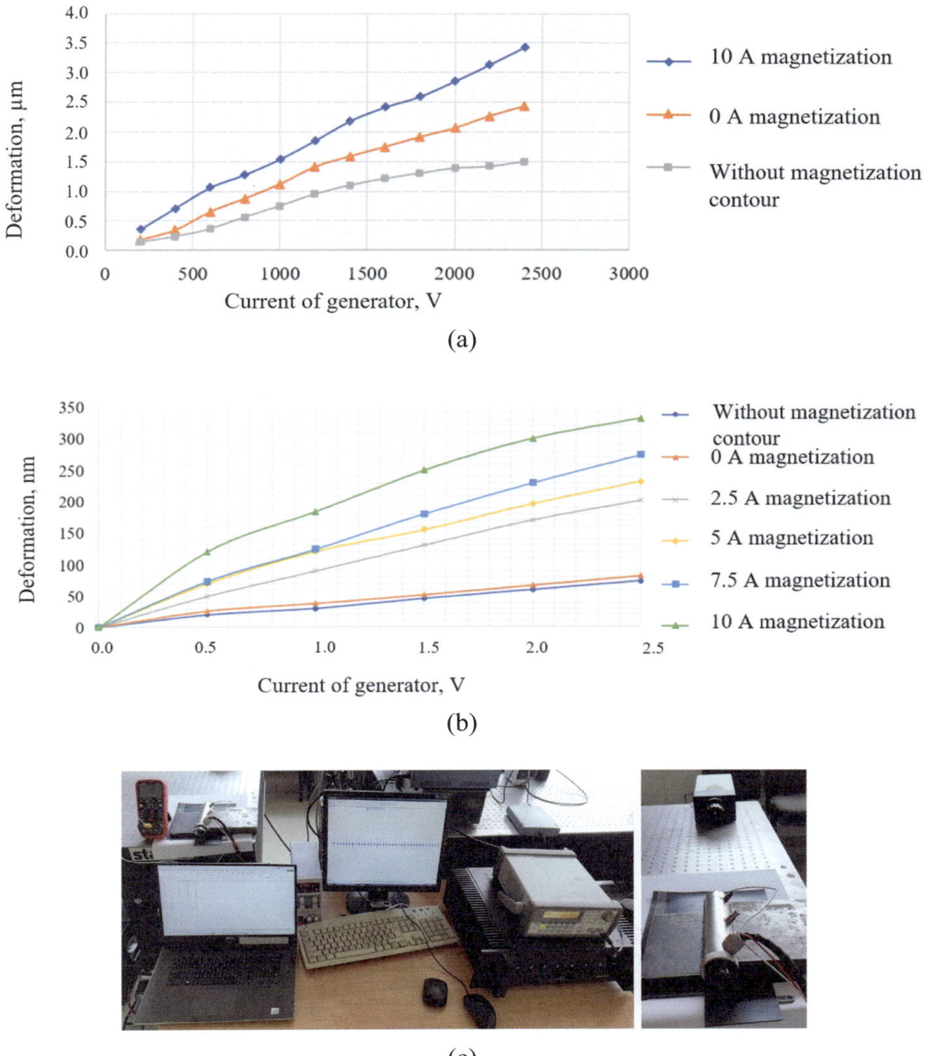

Fig. 3.13 Vibration analysis using a Polytech single-laser system: **a** magnetoconductor oscillation amplitudes; **b** vibration amplitudes of the forming tool; **c** photographs of equipment

References

1. Wu G, Xie P, Yang H, Dang K, Xu Y, Sain M, Turng LS, Yang W (2021) A review of thermoplastic polymer foams for functional applications. J Mater Sci 56:11579–11604. https://doi.org/10.1007/s10853-021-06034-6
2. Li L, Li W, Wang H, Zhao J, Wang Z, Dong M, Han D (2018) Investigation of Prony series model related asphalt mixture properties under different confining pressures. Constr Build Mater 166:147–157. https://doi.org/10.1016/j.conbuildmat.2018.01.120
3. Genovese A, Farroni F, Sakhnevych A (2022) Fractional calculus approach to reproduce material viscoelastic behavior, including the time temperature superposition phenomenon. Polymers 14:4412. https://doi.org/10.3390/polym14204412
4. COMSOL Multiphysics®, v. 6.0; COMSOL AB, Stockholm, Sweden (2024)

Experimental Testing of Functional Prototype of Magnetostrictive Transducer

4

4.1 Pilot Study of Hot Stamping Process

When developing an improved microstructuring technology, it is necessary to compare the formation processes with existing ones to see the influence and changes that occurs with the newer process. For this reason, a magnetostrictive transducer for the hot stamping process was developed, and microstructures of PP, PETG, PVC, and SAN thermoplastics were produced.

The forming equipment consists of a *Tinius Olsen* test machine (*H10KT*, *Tailored Test Solutions Ltd.*, Pennsylvania, USA), a base, a plastic sheet, a heating element, and a controller. The forming equipment is shown in Fig. 4.1.

A nickel matrix was used as the main microstructure. The matrix was heated to the desired temperature and held for a certain time to evenly distribute the temperature throughout the volume of the matrix. The force, temperature, and duration of formation varied. The microstructures were imprinted on four different thermoplastics, PP, PETG, PVC, and SAN. A *Hitachi S-3400N* scanning electron microscope (SEM) (*Hitachi Ltd.*, Tokyo, Japan) was used to study the structures obtained. Images of the structures are presented in Fig. 4.2.

SEM images showed that higher surface quality was obtained with PETG and PP plastics compared to PVC and SAN thermoplastics. To better analyse the quality of the microstructures, the study was performed with an atomic force microscope (AFM) obtaining surface images of topographic profiles. The samples were measured in dynamic contact mode. The results are presented in Figs. 4.3, 4.4, 4.5, 4.6 and 4.7.

The highest quality microstructure formed was obtained in PETG thermoplastic: the average surface roughness was $R_a = 174$ nm, and the average depth of the microstructure was 1000 ± 20 nm. The microstructure of the PP thermoplastic had a smooth surface relief with an average surface roughness of $R_a = 340.2$ nm, and the microstructure depth was

A. Palevicius et al., *Nano/Micro Functional Elements Formation for Bioengineering Applications*, Synthesis Lectures on Biomedical Engineering,
https://doi.org/10.1007/978-3-031-81509-6_4

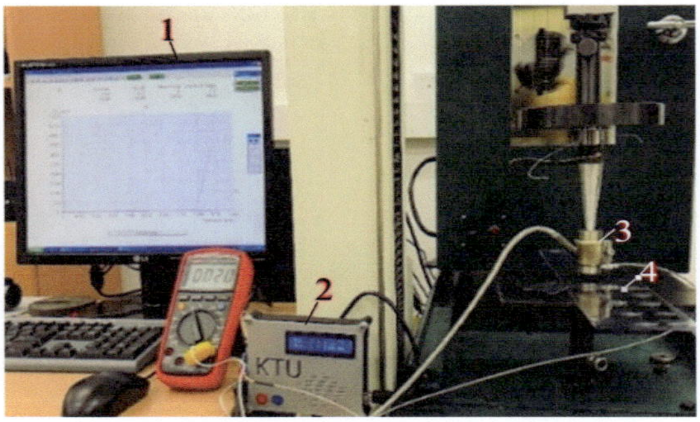

Fig. 4.1 Experimental equipment for microstructure formation: 1—computer, 2—temperature controller, 3—heating element, 4—plastic sheet

similar to the structure formed in PETG plastic, 950 ± 25 nm. The microstructure formed in PVC and SAN thermoplastics had many surface defects with an average roughness of $R_a = 437.6$ nm and $R_a = 298.7$ nm, respectively. From the profile image, the surface shape was not uniform and does not match the basic microstructure.

SEM and AFM measurements showed that the highest quality microstructures were obtained from PP and PETG thermoplastics. PVC and SAN plastics, which had a higher brittleness than PP and PETG plastics, and the thermal properties of the plastics determined the worse reproducibility of the structures.

The quality of the microstructure and the geometrical parameters can determine the optical properties of the structure. Diffraction maxima are obtained by laser illumination through the microstructure. In case of an irregular microstructure or high surface roughness, the diffraction maximum changes. During the study, diffraction efficiency measurements were performed at different maximums. Since PP plastic is not transparent, this material was not investigated in the study. PETG, PVC and SAN microstructures were illuminated with a green light laser with a wavelength of 532 nm. The light maximums were measured by a photodiode and recorded on a computer. The experimental equipment is presented in Fig. 4.8.

One of the main parameters for evaluating the optical properties of the microstructure is the relative diffraction efficiency (RDE), which can be calculated according to the following equation:

$$RDE_{i,j} = \frac{P_{i,j}}{\sum_i P_{i,j}} \tag{4.1}$$

where $RDE_{i,j}$—relative diffraction efficiency; $P_{i,j}$—maximum power of light intensity.

The diffraction efficiencies of three different plastics were theoretically calculated. The following values of the refractive indices of plastics were used in the calculations: PETG—1.57 [1], PVC—1.531 [2], SAN—1.572 [3]. The theoretical influence of the depth of the microstructure on the diffraction efficiency was calculated using refractive indices. Without evaluating the change in the quality of the surface, the depth of the

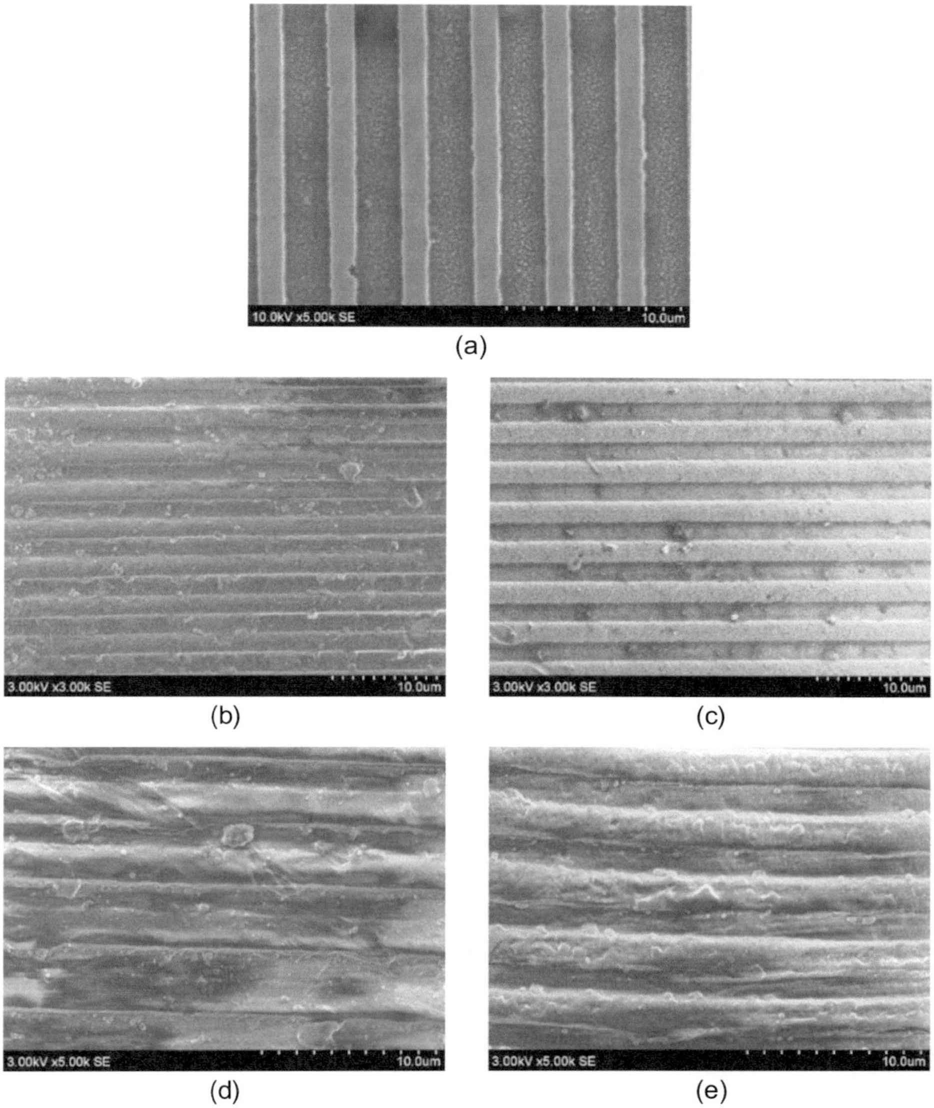

Fig. 4.2 SEM images of microstructures: **a** image of nickel microstructure; **b** PP thermoplastic; **c** PETG thermoplastic; **d** PVC thermoplastic; **e** SAN thermoplastic

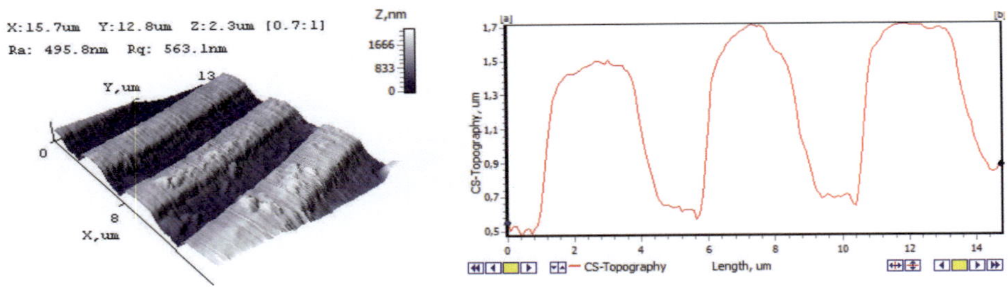

(a) (b)

Fig. 4.3 AFM topography and surface profile images of nickel microstructure

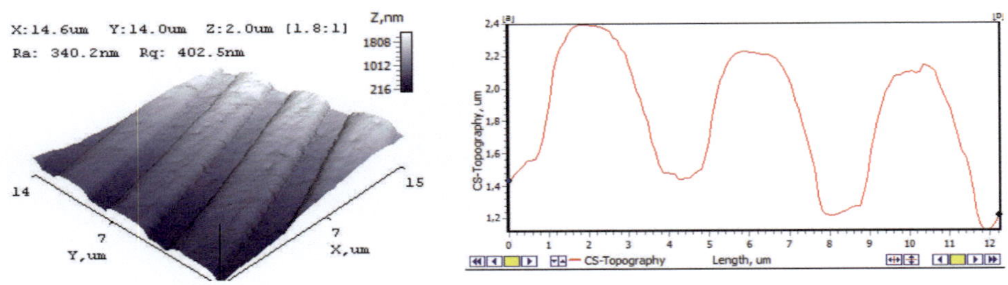

(a) (b)

Fig. 4.4 AFM topography and surface profile images of PP thermoplastic

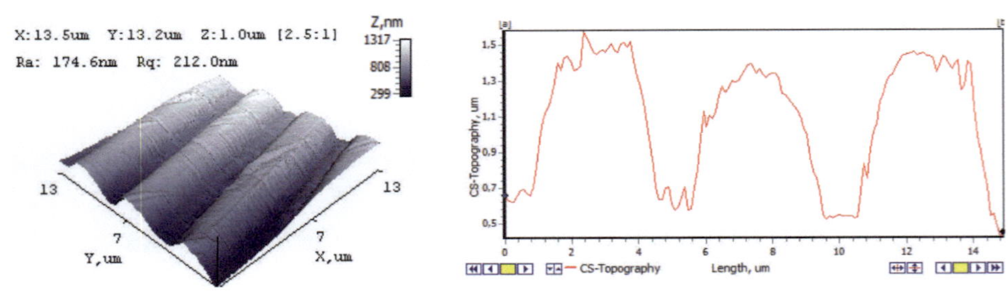

(a) (b)

Fig. 4.5 AFM topography and surface profile images of PETG thermoplastic

formed microstructure can be determined on the basis of theoretical calculations and measurements of the diffraction maximum. The theoretical graph of the maximum diffraction efficiency is presented in Fig. 4.9.

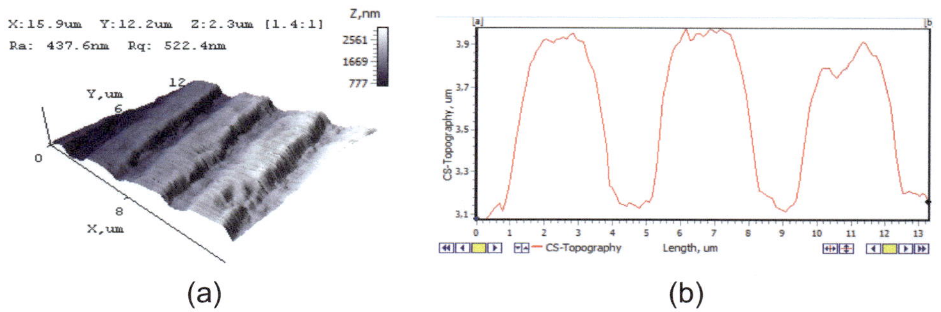

Fig. 4.6 AFM topography and surface profile images of PVC thermoplastic

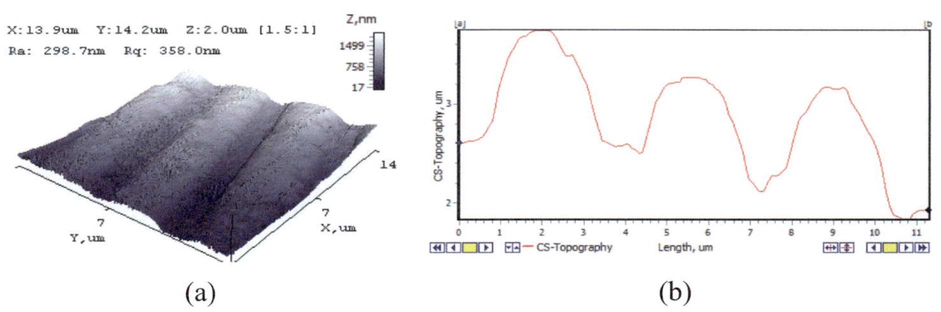

Fig. 4.7 AFM topography and surface profile images of SAN thermoplastic

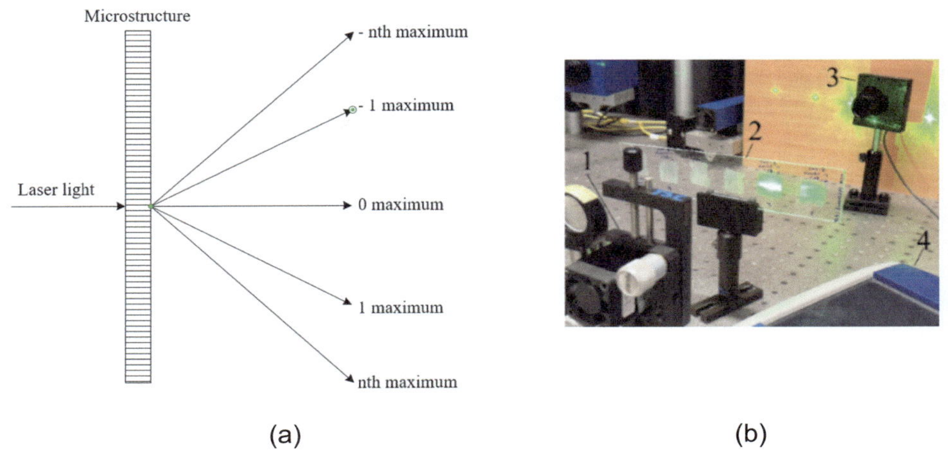

Fig. 4.8 Measurement of diffraction efficiency: **a** formation of diffraction maxima; **b** equipment: 1—laser, 2—matrix, 3—photodiode, 4—display screen

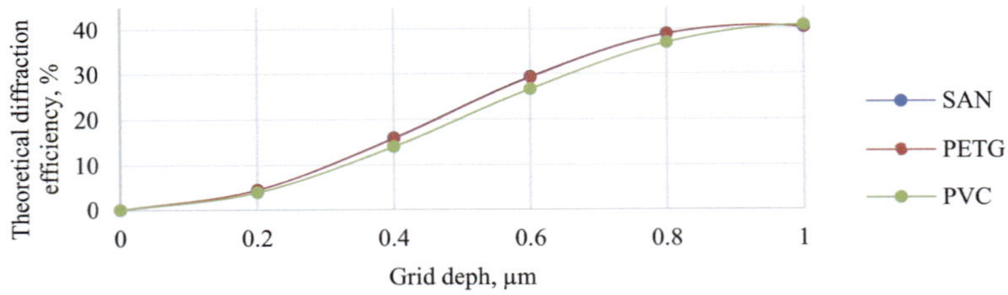

Fig. 4.9 Theoretical dependence of the diffraction efficiency of thermoplastics on the depth of the microstructure

Twelve diffraction efficiency tests were performed for each thermoplastic in various measurement settings (temperature, stress, and duration of the indentation).

When the PETG microstructure was formed with a load of 2000 N, a molding duration of 10 s, and a temperature of 125 °C, the best relative diffraction efficiency, or RDE = 34.62%, was obtained. When the indentation settings were set to 5000 N load, 10 s of formation time, and 130 °C temperature, the RDE of the structures created in SAN plastic was 29.04%. Moreover, RDE = 22.44% was observed in PVC plastic microstructures after applying 5000 N force for 10 s at 80 °C. The results of the diffraction efficiency measurements are presented in Tables 4.1, 4.2 and 4.3.

When the molding temperature was greater than the plastic glass transition temperature, all the formed polymers had higher diffraction efficiencies. The plastic could flow slowly into the mold and stopped further deforming in areas where it was not necessary when the temperature was near the glass transition point. The results demonstrate that the results were also impacted by the formation time.

Table 4.1 PETG diffraction efficiency measurement results

Load, N	Time, s	Temperature, °C	RDE, %
4000	10	100	14.31
5000	5	100	21.43
5000	10	100	20.50
4000	5	100	13.49
2000	10	125	22.62
2000	10	100	18.81
2000	5	100	22.31
3000	10	100	10.67
2000	10	90	14.85
2000	5	90	19.86
2000	10	80	13.54

Table 4.2 PVC diffraction efficiency measurement results

Load, N	Time, s	Temperature, °C	RDE, %
5000	10	100	5.39
5000	5	100	12.63
4000	10	100	7.28
4000	5	100	11.59
3000	10	100	5.43
3000	5	100	11.66
4000	10	125	10.44
4000	5	125	7.38
5000	5	90	6.03
5000	10	80	22.44
2000	10	125	6.74
2000	5	125	6.65

Table 4.3 SAN diffraction efficiency measurement results

Load, N	Time, s	Temperature, °C	RDE, %
5000	10	100	17.07
5000	10	120	16.49
5000	15	120	14.94
5000	10	140	12.52
4000	10	140	13.28
4000	5	140	7.85
3000	10	140	18.12
3000	5	140	16.85
3000	10	130	18.23
5000	2	130	17.76
5000	10	130	19.04
4000	10	130	15.25

4.2 Study of the Hot Stamping Process with a Magnetostrictive Transducer

Heating, embossing, and cooling are the three main phases of the hot embossing process. Plastic is usually heated to the glass transition temperature, embossed, the load is removed after a specific period, and the produced structure is given time to cool. The plastic heating process and the embossing step were combined using the improved molding tool [4]. In addition to applying vibration throughout the molding process, the mold was heated to

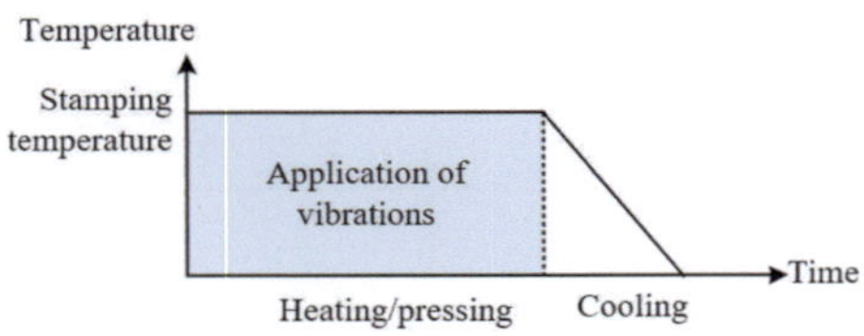

(a)

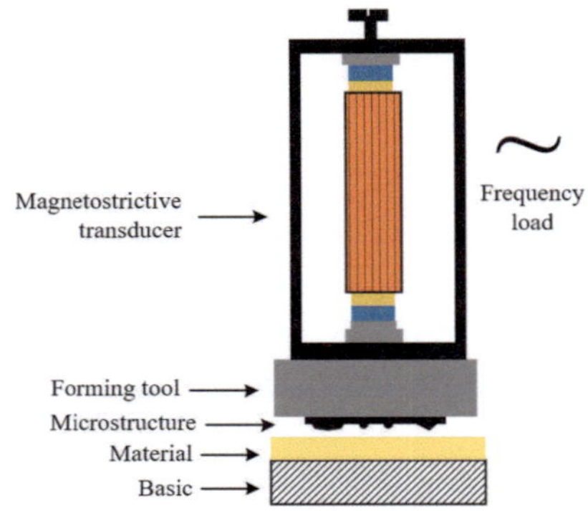

(b)

Fig. 4.10 Hot stamping process with a magnetostrictive transducer: **a** temperature variation curve during the hot stamping procedure using a magnetostrictive transducer; **b** principle of operation of the magnetostrictor

the specified temperature. The graph of the formation stages and visualization of the formation process are presented in Fig. 4.10.

The forming equipment consisted of a magnetostrictor, a forming microstructure, a polymer plate, and a base. The test was performed using a *Tinius Olsen* tensile testing machine (*H10KT*, *Tailored Test Solutions Ltd.*, Pennsylvania, USA). Additional components used in the system: signal generator (*UTG2025A*, UNI-T, China), amplifier (*PA8HF*, Wilcoxon, England) and power supply (*KPS3010D*, Wanptek, China). With this equipment, it was possible to achieve a maximum deformation of 4 μm. The temperature of the forming tool was recorded with a digital thermometer (*UT161D*, UNI-T, China). The experimental equipment is presented in Fig. 4.11.

The forming equipment was selected to be able to vary the forming force, temperature, vibration amplitude, and frequency. Imprints were repeated three times to ensure the reliability and accuracy of the data. The change in structure quality was assessed by measuring the change in diffraction efficiency. A green (525 nm) light laser (*OX-MZ5201*, OXLasers, China) and a high-power laser light detector with a monitor (*11UP12*, Gentec, Canada) were used for the optical measurements. A computer (*Inspiron 7560*, Dell, USA) was used for additional data recording, and stands were used to hold the formed microstructure and the laser. Measurement equipment is presented in Fig. 4.12.

Diffraction measurement results are presented in Figs. 4.13, 4.14 and 4.15.

The diffraction gratings were created in samples of three different thermoplastics that were measured for diffraction research. Average results of three identical gratings were used to calculate the diffraction efficiency. The optimal conditions for the formation of a diffraction grating in PVC plastic were determined to be 100 °C, 10 kHz excitation frequency, 2 V excitation voltage, 10 A magnetization current and 3 kN load. This was

Fig. 4.11 Microstructure formation equipment: 1—tensile testing machine, 2—magnetostrictive transducer, 3—microstructure, 4—amplifier, 5—signal generator, 6—power source, 7—thermometer

Fig. 4.12 Diffraction measurement equipment: 1—laser, 2—laser light detector, 3—monitor, 4—microstructure, 5—computer

determined by measuring the diffraction efficiency. The best RDE was achieved in PETG plastic at 170 °C, 10 kHz excitation frequency, 2 V excitation voltage, 10 A magnetization current, and 1 kN load. Regarding the SAN plastic, the optimal RDE was achieved at 110 °C, 10 kHz excitation frequency, 2 V excitation voltage, 0 A magnetization current, and 1 kN load. The results showed that the diffraction efficiency increased by 18.09% for PVC plastic, 2.92% for PETG plastic, and 21.32% for SAN plastic compared to the non-vibration molding process. Based on theoretical calculations, this magnetostrictive transducer has been shown to be superior to other currently available analogous devices because of its increased diffraction efficiency and decreased residual stresses.

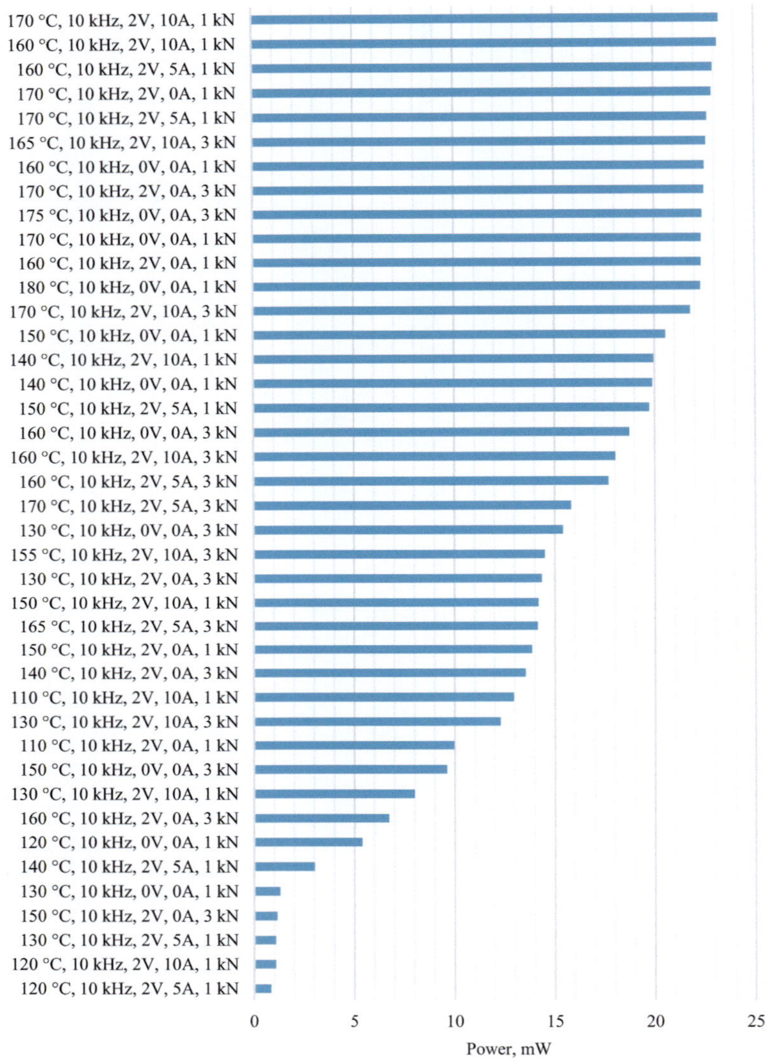

Fig. 4.13 Measurement results of diffraction efficiency of microstructures formed in PETG plastic

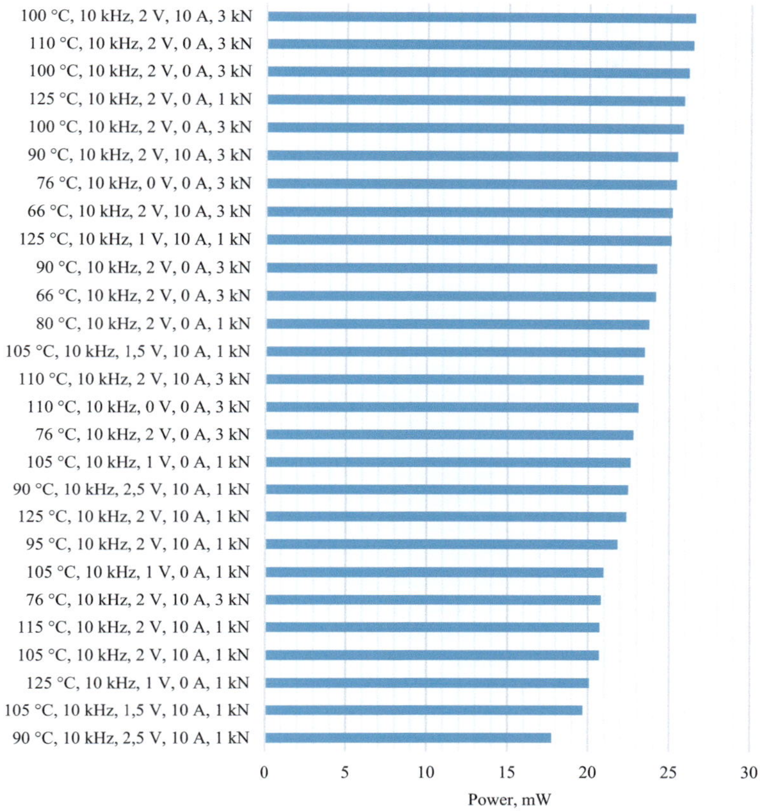

Fig. 4.14 Measurement results of diffraction efficiency of microstructures formed in PVC plastic

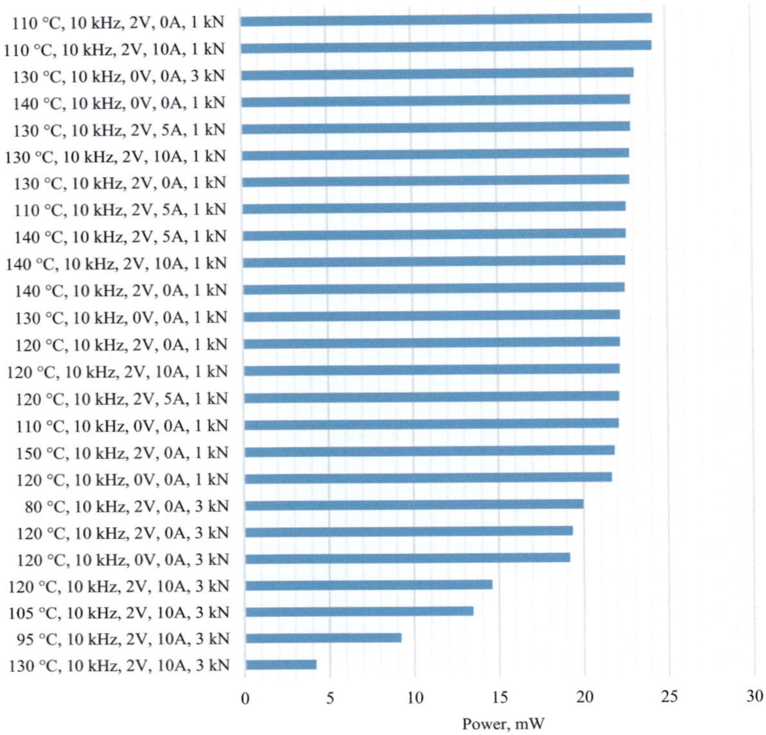

Fig. 4.15 Measurement results of diffraction efficiency of microstructures formed in SAN plastic

References

1. Fohtung E (2022) Magnetostriction fundamentals. In: Encyclopedia of smart materials, vol 4, pp 130–133. https://doi.org/10.1016/B978-0-12-815732-9.00081-4
2. Mariani A, Malucelli G (2023) Insights into induction heating processes for polymeric materials: an overview of the mechanisms and current applications. Energies 16:4535. https://doi.org/10.3390/en16114535
3. Martins P, Fernandez CSL, Silva D, Lanceros-Méndez S (2021) Theoretical optimization of magnetoelectric multilayer laminates. Compos Sci Technol 204:108642. https://doi.org/10.1016/j.compscitech.2020.108642
4. Ciganas J, Bubulis A, Jurenas V, Griskevicius P, Palevicius A, Urbaite S, Janusas G (2023) Dynamic mechanical properties of PVC plastics in the formation of microstructures with novel magnetostrictor. Micromachines 14:820. https://doi.org/10.3390/mi14040820

5 Development and Analysis of Electrochemical Reactor with Vibrating Functional Element for Nanoporous AAO Membranes Fabrication

5.1 Development of Electrochemical Reactor with Vibrating Functional Element

Since it is important to properly regulate the geometry of the nanostructure, the electrochemical anodization method is used to obtain precise geometrical properties of nanoporous AAO membranes. Although the anodization process is well known and studied, the production of nanoporous AAO membranes is a complicated process in which self-changing factors influence the final membrane geometry. For example, alumina's growth rate varies during the anodization process, even when growth occurs at constant voltage and electrolyte concentration. It is well known that the AAO growth rate is temperature dependent. As a result, higher temperatures promote growth, whereas lower temperatures slow growth [1]. However, it is important to note that the concentration of the electrolyte changes gradually during the anodization process, therefore any change in the composition and concentration of the electrolyte necessarily impacts the anodization rate [2–4]. All these complicating aspects make accurate control of the anodization process challenging, especially when the final membrane thickness requirements are high [5, 6]. As a result, a temperature-controlled electrochemical reactor is being developed that maintains the electrolyte at a consistent temperature during the anodization process.

The use of high-frequency excitation during the anodization process to control the AAO membrane nanopore geometries is in the early stages of research. The finding that vibrations affect the membrane pore diameter encourages the development and analysis of new fabrication technologies for controlling the geometry of nanoporous AAO membranes [7]. For that purpose, the temperature-controlled electrochemical reactor is supplemented with a piezoelectric ceramic material that generates high-frequency oscillations during excitation. After studying the theoretical growth models of AAO, it was found that the concentration and temperature of the electrolyte changes along the nanopore. Based on

A. Palevicius et al., *Nano/Micro Functional Elements Formation for Bioengineering Applications*, Synthesis Lectures on Biomedical Engineering,
https://doi.org/10.1007/978-3-031-81509-6_5

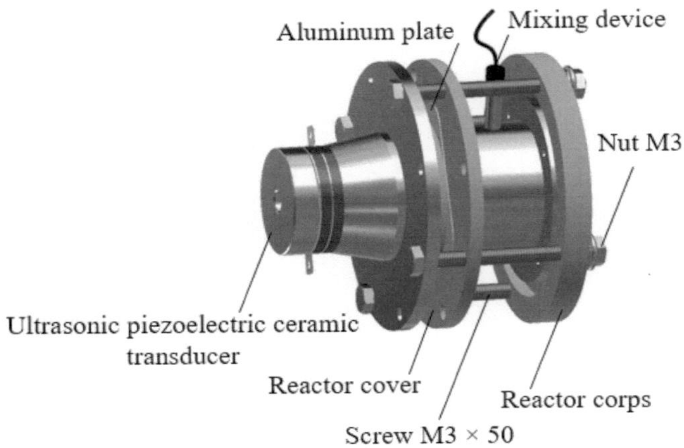

Fig. 5.1 3D model of the novel design of a temperature-controlled electrochemical reactor with vibrating functional element

this; to control the pore geometry, it is necessary to control the changes in temperature and pH values during anodization. For example, studies have shown that the laminar mixing efficiency of hot and cold laminar fluid flows can be improved by using acoustic flow [8]. Surface acoustic waves apply the effect of ultrasonic waves and have gained attention due to their advantages in effective fluid control and their non-invasive nature [9]. High-frequency oscillations contribute to variations in temperature and pH values in the nanopores of the AAO membrane during the formation of the oxide layer, thereby ensuring the electrolyte mixing process and the renewal of flow along the entire length of the pore [10]. The 3D model of the novel design of the temperature-controlled electrochemical reactor with vibrating functional element is presented in Fig. 5.1.

The temperature-controlled electrochemical reactor with vibrating functional element has been developed and its structure is shown in Fig. 5.2.

Since acid is used during anodization, the reactor corps is made of stainless steel AISI 304. In addition, other components of the reactor, such as the cover, screws, nuts, and impeller, are made of stainless steel. A four-bladed mixing impeller with a diameter of 20 mm, a height of 10 mm, blade angle of 0° and rotational speed of 108 rpm is used for mixing the electrolyte during the anodization process. Since the electrolyte temperature of 5 °C is a critical parameter of the anodization process to ensure the formation of nanoscale pores, a temperature sensor is placed in the assembled reactor to monitor and control the temperature of the electrolyte. The Peltier element is used to cool the electrolyte, and the cooler is connected to the Peltier element and the reactor corps. The Peltier element is lubricated on both sides with thermal paste, which improves the temperature exchange and increases the contact area between the Peltier element, the cooler, and the reactor corps.

Fig. 5.2 The temperature-controlled electrochemical reactor with vibrating functional element

To produce the nanoporous AAO membrane, aluminum is the main material. A 0.5 mm thick aluminum plate is attached to the stainless-steel reactor cylinder. An acid-resistant rubber gasket is used to prevent liquid leakage. The rubber gasket is installed between the aluminum plate and the reactor cylinder. To generate high-frequency excitation, an ultrasonic piezoelectric ceramic transducer is attached to the other side of the aluminum plate. All reactor components are assembled by fittings. Screws with dimensions M3 × 50 and nuts are used to support the structure. This reinforcement guarantees good contact and vibration transmission during the anodization process.

Once the main electrochemical reactor is developed, all the equipment needed for the successful production of nanoporous AAO membranes needs to be prepared and assembled. For that purpose, an experimental stand has been prepared, which is presented in Fig. 5.3.

The experimental stand consists of a 12 V and 15 A direct current power supply unit, a 60 V and 5 A direct current power supply unit, a master cooler radiator, automatic temperature control system and the novel design of the electrochemical reactor. The automatic temperature control system is composed of temperature control device, temperature sensor, and Peltier element.

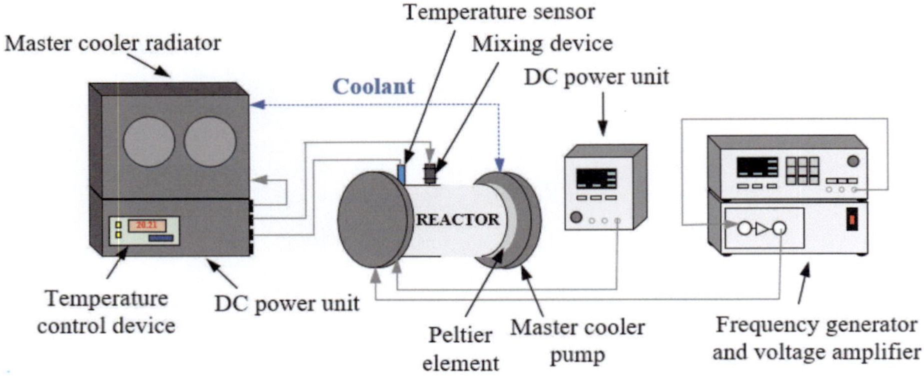

Fig. 5.3 A schematic representation of the experimental setup

5.2 Analysis of Electrochemical Reactor with Vibrating Functional Element

The designed temperature-controlled electrochemical reactor was tested to evaluate its suitability to produce nanoporous AAO membranes during the two-step anodization process.

To make the electrolyte temperature and concentration the same throughout the reactor, a four-bladed mechanical impeller was used. Reactor mixing process experiments were conducted to monitor whether the impeller sufficiently mixed the electrolyte.

Initially, the modelling method was chosen to analyse the mixing process. Theoretical simulation of the mixing process was performed using *Ansys 17* software (*Ansys*®, Pennsylvania, United States). The simulation model of the mixing process with the coordinate system is presented in Fig. 5.4.

The model was created using the actual dimensions of the electrochemical reactor and impeller. The conditions of the analysis were chosen to be approximate to the real conditions. The simulation was performed at a temperature of 20 °C. Water was chosen as the liquid. The water density was 998.2 kg/m^3, viscosity was 0.001003 kg/m s. Furthermore, the force of gravity was included in the calculations. The theoretical velocity vectors obtained in different planes are shown in Fig. 5.5.

Theoretical results showed that the entire volume of the electrolyte liquid in the reactor is mixed. Furthermore, after the theoretical simulations, the experimental study was carried out. A transparent prototype reactor of a real-size reactor was fabricated to conduct an experimental mixing study. The mixing experiment used water and thermoplastic polymer pellets of acrylonitrile butadiene styrene (ABS) with a density of 1032–1380 kg/m^3 to visualize particle movement [11]. When the mixing process started, the settled plastic

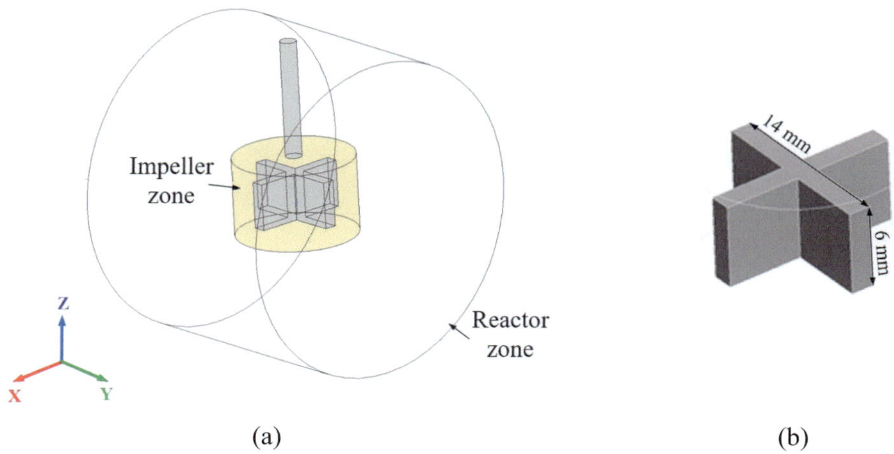

Fig. 5.4 Simulation model of the mixing process: **a** model with coordinate system; **b** size of impeller

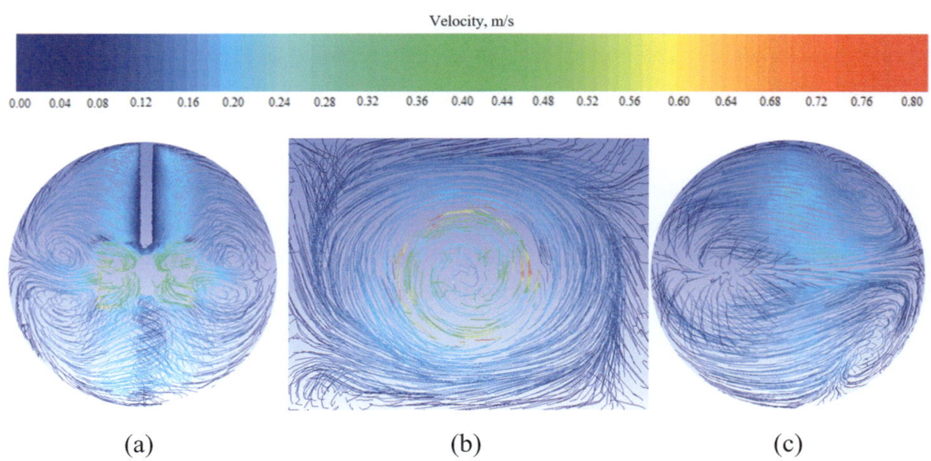

Fig. 5.5 Velocity vectors in different planes: **a** x–z; **b** x–y; **c** x–y–z

particles started to move and distribute throughout the reactor volume. Images from the experiment are presented in Fig. 5.6.

Analysis of the mixing process assumes that the mixing process was sufficient to mix the electrolyte inside the reactor during the anodization process.

In addition, to evaluate the temperature of the reactor electrolyte, a temperature variation experiment was performed. The automatic temperature control system was used.

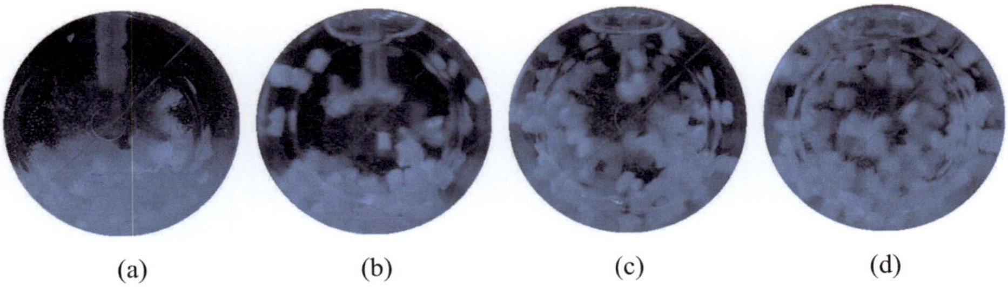

Fig. 5.6 Experiment of the mixing process: **a** start before the mixing process; **b** 0.5 s after the mixing process; **c** 1 s after the mixing process; **d** 1.5 s after the mixing process

The temperature sensor was attached near the aluminum plate to monitor the temperature changes of the electrolyte during the anodization process. When the temperature sensor detected an electrolyte temperature greater than 5 °C, the temperature control device automatically turned on the Peltier element, which cooled the liquid in the reactor. When the liquid reached a temperature of 5 °C, the Peltier element was automatically switched off. The process was repeated while the anodization process was in progress. Electrolyte temperature analysis was performed in two ways: (1) the impeller was turned off; (2) the impeller was turned on. The results of analysis are presented in Fig. 5.7.

During analysis, when the impeller was turned off, it was experimentally determined that the temperature sensor near the aluminum plate did not reach the required temperature of 5 °C in 600 s. The automatic control system has been continuously activated, but the temperature was not uniform throughout the reactor as a result of the nonstirring liquid. An ice has formed near the Peltier element. The formation of ice confirms the formation of different temperature zones. In the second experiment, when the impeller was turned on, the required temperature of 5 °C was achieved. The temperature of the electrolyte

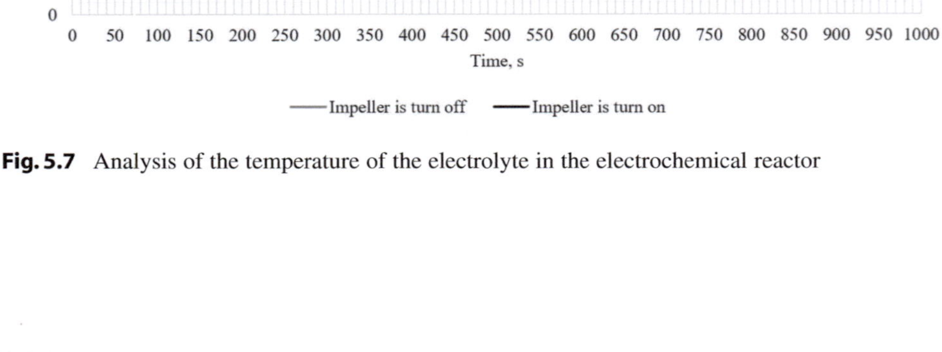

Fig. 5.7 Analysis of the temperature of the electrolyte in the electrochemical reactor

inside the reactor varied from 4.5 to 5.5 °C with the automatic temperature control system. Temperature changes of one degree are permissible during the anodization process. The automatic temperature control system was found to be suitable for reactor design. Moreover, the automatic temperature control system can be used not only to maintain a temperature of 5 °C. The system can maintain temperatures from − 10 to + 70 °C. Therefore, the novel design of the reactor can be used for other processes where temperature control is required.

After the suitability of the reactor temperature and the mixing process was determined, the vibration modes of the aluminum plate were determined.

To control the pore geometry of the nanoporous AAO membrane using vibration, it is important to ensure that the aluminum plate vibrates at high-frequency and the vibration modes of the aluminum plate are obtained. For this purpose, theoretical and experimental vibration analysis was performed. First, vibration analysis was performed by modelling aluminum plates vibration modes with *COMSOL Multiphysics 5.4* software (*COMSOL®*, Burlington, United States). The aluminum plate with a diameter of 40 mm and a thickness of 0.5 mm was used for the simulation. The drawing of a simulation model of the vibration process is presented in Fig. 5.8.

In the simulation model, aluminum was the main material. Aluminum Young's modulus was 68 GPa, the mass density was 2712 kilogrammes per cubic metre (kg/m^3), material Poisson's ratio was 0.33, pretension force was 19,000 Newtons per metre (N/m). The finite tetrahedron element method was used in the simulation. The effect of vibration on the aluminum plate is shown in Table 5.1.

Mode shapes can be written with two numbers. In the descriptions of the circular membrane mode shapes, the first number indicates the number of nodal diameters, and the second number indicates the number of nodal circles.

After the results of the theoretical simulation of the five modal shapes were obtained, an experimental study was performed. A noncontact holographic measurement system (the precise real-time instrument for surface measurement, also known as the *PRISM* system

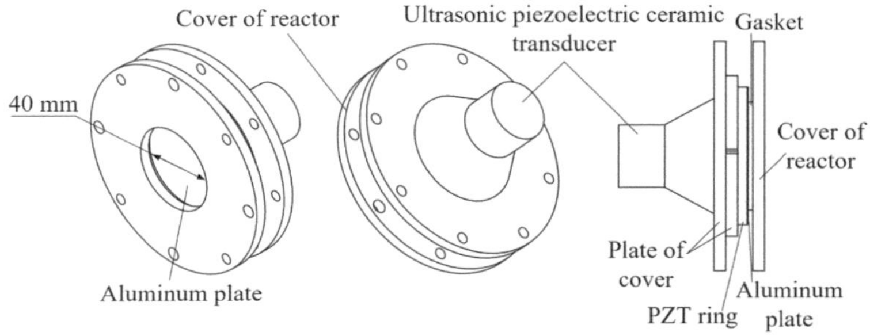

Fig. 5.8 Simulation model of the vibration process

Table 5.1 Surface displacement field at different mode shapes

Mode shapes	Simulation results	Frequency (kHz)
(0,1)	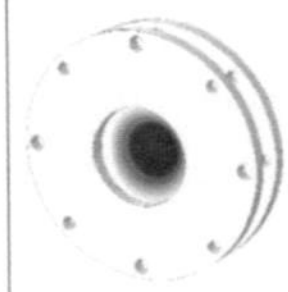	3.0
(1,1)	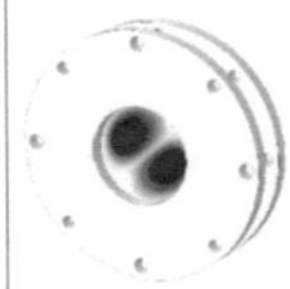	4.8
(2,1)	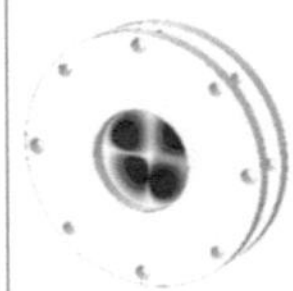	6.5
(0,2)	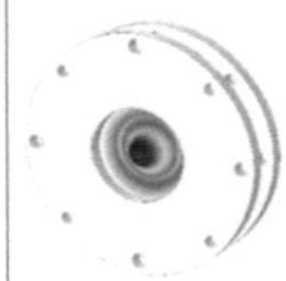	6.9
(3,1)	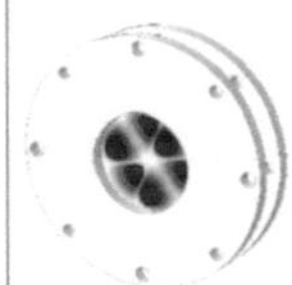	8.8

(*Hytec*, Los Alamos, United States)) was used for experimental vibration measurement. The experimental setup is presented in Fig. 5.9.

Green laser light with a wavelength of 532 nm was scattered from the object (aluminum plate) and collected by the camera lens. The captured information is transferred to a computer where the *PRISM DAQ* programme (*Hy-Tech Forming Systems*, Los Alamos, USA) was used for data analysis. The obtained experimental results of the vibration modes of the aluminum plate are summarized in Table 5.2.

The analysis of the temperature controlled electrochemical reactor was performed. For high-frequency excitation of the aluminum plate, the vibrating functional element was attached to the reactor. After analysis, it was found that the electrochemical reactor meets the requirements of mixing, temperature, and vibration processes, which are important during the anodization process. In further research, nanoporous AAO membranes may be produced using the novel design of an electrochemical reactor.

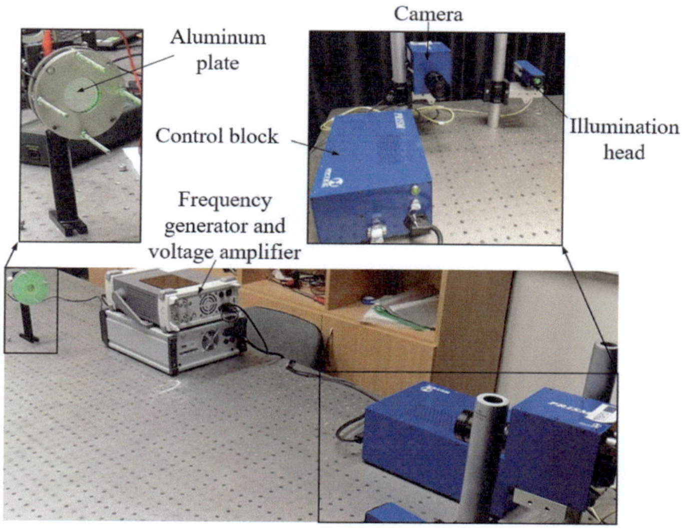

Fig. 5.9 PRISM system for vibration experiments

5.3 Nanoporous AAO Membranes Fabrication Using Electrochemical Reactor with Vibrating Functional Element

An electrochemical reactor with a vibrating functional element is also used to produce nanoporous AAO membranes. For this purpose, a two-step anodization process is used. The production stages are shown in Fig. 5.10.

For the experiments, 0.5 mm thick and 5 × 5 cm square aluminum plates (1050 A, 99.5%) were used. Aluminum plates must be annealed at 400 °C for 4 h in a nitrogen atmosphere in a conventional furnace. Before the first step of anodization, the samples were degreased with acetone and rinsed with distilled water. The prepared samples were anodized in an electrochemical reactor at a voltage of 60 V and a temperature of 5 °C for one hour. The electrolyte was 0.3 M oxalic acid ($H_2C_2O_4$). After the first step, the oxide layer was chemically exposed to a 3.5% concentrated phosphoric acid (H_3PO_4) and 2% chromic anhydride (CrO_3) acid solution in water at 80 °C for 10 min and washed with distilled water. The second anodization step was carried out at 60 V and 5 °C for eight hours. Finally, the samples were washed with distilled water and air-dried.

Using the same method as used for producing porous AAO membranes, high-frequency vibrations were used. To determine the effect of vibrations on the membrane pore geometry, it was decided to conduct the study using the first vibration mode of aluminum plates. High-frequency excitation was applied to the aluminum plate during the second-step of the anodization process. 20 nanoporous AAO membranes were produced during

Table 5.2 Results of the theoretical and experimental analysis of the vibration process

Simulation results	Theoretical frequency (kHz)	Experimental results	Experimental frequency (kHz)
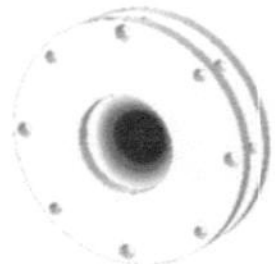	3.0		3.4
	4.8		4.3
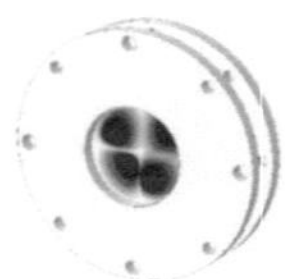	6.5	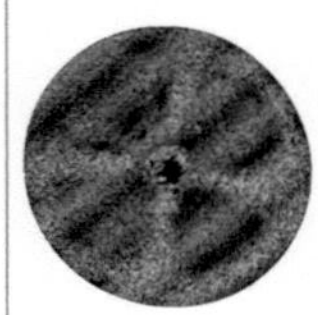	6.4
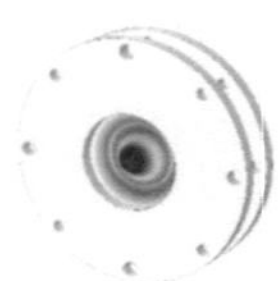	6.9	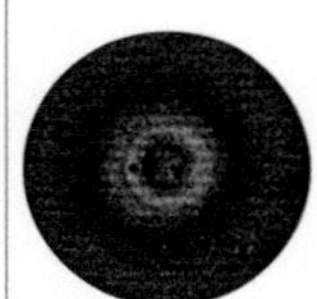	6.8
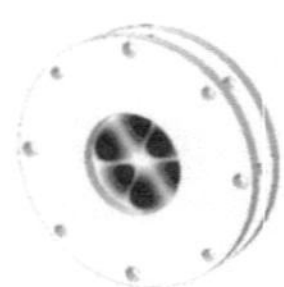	8.8	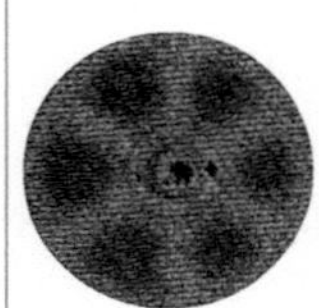	9.1

the experiment (10 samples without vibration and 10 samples when the aluminum plate reached the first resonant mode). The morphology of the nanoporous AAO membranes was determined by *Hitachi S-3400N* SEM (*Hitachi Ltd*, Tokyo, Japan). The pore diameter (D_p) and interpore distance (D_c) were determined using *ImageJ* software (*National Institutes of Health*, JAV). Based on the diameters of the pores and the distance between the pores, the porosity (P) of the membrane was calculated according to the formula [12]:

$$P = 0.907 \cdot \left(\frac{D_p}{D_c}\right)^2 \% \tag{5.1}$$

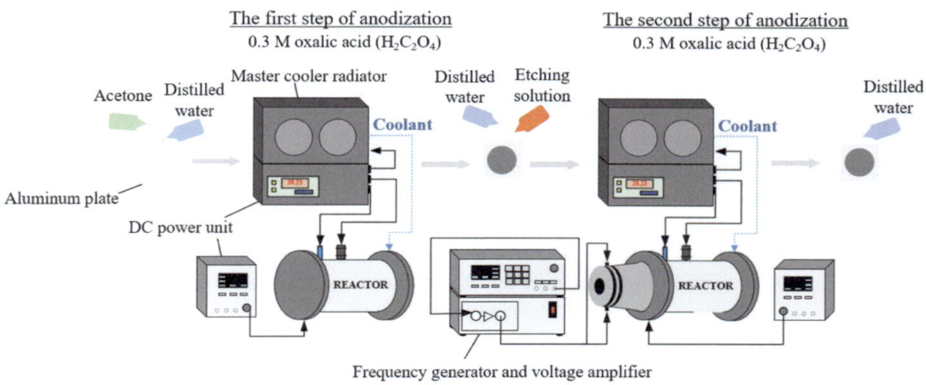

Fig. 5.10 The stages of the two-step anodization process

Table 5.3 Geometric parameters (D_p, D_c, and P) of the nanoporous AAO membranes

Parameter	No frequency excitation	Frequency excitation at 3.4 kHz
Pore diameter (nm)	57.0 ± 10	84.5 ± 10
Interpore distance (nm)	120.0 ± 20	121.7 ± 20
Porosity (%)	20	44

where D_p—pore diameter; D_c—interpore distance.

Porosity can be defined as the ratio of the surface area occupied by pores to the total surface area. The geometric parameters (D_p, D_c, and P) of nanoporous AAO membranes are presented in Table 5.3.

The analysis showed that the use of the high-frequency excitation method during the two-step anodization process changes the pore geometry. The change in the pore size affect the porosity parameter of the AAO membrane, which provides a real opportunity to create functional nanoporous AAO membranes that could be applied in different fields. Such results encourage more extensive research on nanoporous AAO membranes. Therefore, the next section presents the influence of high-frequency excitation on the characteristics of the nanoporous AAO membrane when vibrations are applied to the aluminum plate during the two-step anodization process.

References

1. Araujo JVS, Milagre M, Costa I (2023) A historical, statistical and electrochemical approach on the effect of microstructure in the anodizing of Al alloys: a review. Crit Rev Solid State Mater Sci 521–581. https://doi.org/10.1080/10408436.2023.2230250
2. Gasco-Owens A, Veysrenaux D, Cartigny V, Rocca E (2021) Large pores anodizing of 5657 aluminum alloy in phosphoric acid: an in-situ electrochemical study. Electrochim Acta 382:138303. https://doi.org/10.1016/j.electacta.2021.138303
3. Terashima A, Iwai M, Kikuchi T (2022) Nanomorphological changes of anodic aluminum oxide fabricated by anodizing in various phosphate solutions over a wide pH range. Appl Surf Sci 605:154687. https://doi.org/10.1016/j.apsusc.2022.154687
4. Poznyak A, Pligovka A, Laryn T, Salerno M (2021) Porous alumina films fabricated by reduced temperature sulfuric acid anodizing: morphology composition and volumetric growth. Materials 14(4):767. https://doi.org/10.3390/ma14040767
5. Wang K, Cao Y, Cui Y, Ye A, Yi S, Hu Z (2022) Study on parameter correlation of thickness and performance of anodizing film on 6061 aluminum alloy frame in high energy laser system. Coatings 12(12):1978. https://doi.org/10.3390/coatings12121978
6. Bruera FA, Kramer GR, Vera ML, Ares AE (2020) Evaluation of the influence of synthesis conditions on the morphology of nanostructured anodic aluminum oxide coatings on Al 1050. Surf Interfaces 18:100448. https://doi.org/10.1016/j.surfin.2020.100448
7. Cigane U, Palevicius A, Jurenas V, Pilkauskas K, Janusas G (2022) Development and analysis of electrochemical reactor with vibrating functional element for AAO nanoporous membranes fabrication. Sensors 22:8856. https://doi.org/10.3390/s22228856
8. Hsu JC, Chang CY (2022) Enhanced acoustofluidic mixing in a semicircular microchannel using plate mode coupling in a surface acoustic wave device. Sens Actuators A Phys 336:113401. https://doi.org/10.1016/j.sna.2022.113401
9. Chen Z, Shen L, Zhao X, Chen H, Xiao Y, Zhang Y, Yang X, Zhang J, Wei J, Hao N (2022) Acoustofluidic micromixers: from rational design to lab-on-a-chip applications. Appl Mater Today 26:101356. https://doi.org/10.1016/j.apmt.2021.101356
10. Maramizonouz S, Jia C, Rahmati M, Zheng T, Liu Q, Torun H, Wu Q, Fu YQ (2022) Acoustofluidic patterning inside capillary tubes using standing surface acoustic waves. Int J Mech Sci 214:106893. https://doi.org/10.1016/j.ijmecsci.2021.106893
11. Xavior M, Nishanth D, Kumar N, Jeyapandiarajan P (2020) Synthesis and testing of FGM made of ABS plastic material. Mater Today Proc 22:1838–1844. https://doi.org/10.1016/j.matpr.2020.03.018
12. Kozhukhova AE, Preez SP, Bessarabov DG (2019) Preparation of anodized aluminium oxide at high temperatures using low purity aluminium (Al6082). Surf Coat Technol 378:124970. https://doi.org/10.1016/j.surfcoat.2019.124970

Vibration-Assisted Synthesis of Nanoporous Anodic Aluminum Oxide (AAO) Membranes

6

Introduction. Nanoporous AAO membranes have become a common template for the preparation of functional nanomaterials due to the simple preparation process, inexpensive materials, controllable nanostructure, corrosion resistance, thermal stability, and mechanical stability [1, 2]. The pore size and aspect ratio of the AAO template can be accurately controlled by modifying manufacturing parameters, such as voltage, current, time, and electrolyte composition [3–5]. This tunability allows the template to be tailored to specific application needs, giving precise control over the size, shape, and dimensions of the produced nanostructures [6, 7]. Nanoporous AAO membranes have been studied for many years and continue to arouse curiosity among many researchers due to their unique chemical and physical properties [8]. Therefore, research related to controlling the geometric properties of nanoporous AAO membranes is relevant. This chapter presents the effect of vibrations on nanoporous AAO membrane pore diameter, interpore distance, porosity, membrane thickness, and chemical composition when the aluminum plate is excited at high-frequency during the two-step anodization process.

6.1 Effect of the Resonance Frequency on the Formation of the AAO Nanopore Structure

Nanoporous AAO membranes were fabricated using the electrochemical reactor with a vibrating functional element to vibrate an aluminum plate during the two-step anodization process. The principle of the production process is presented in Sect. 5.3. According to the findings, aluminum anodization technology, which uses high-frequency excitation in

A. Palevicius et al., *Nano/Micro Functional Elements Formation for Bioengineering Applications*, Synthesis Lectures on Biomedical Engineering,
https://doi.org/10.1007/978-3-031-81509-6_6

the two-step anodization process, provides an opportunity to control the membrane pore diameter by using and changing the resonance frequency of the aluminum plate.

To determine the influence of the resonance frequency on the formation of the AAO nanopore structure, the aluminum plate was vibrated in the first and second vibration modes. During the anodization process, a low-mass accelerometer attached to vibrating aluminum plates was used to monitor vibration modes.

Based on the theoretical results of the vibration analysis, the resonant frequencies of the first and second modes of aluminum plates were selected. After the production process of nanoporous AAO membranes, the results were analyzed on the basis of the SEM images of the membranes, which were processed using *ImageJ* software (*National Institutes of Health*, JAV). 30 nanoporous AAO membranes were produced during the experiment (10 samples without vibration, 10 samples when the aluminum plate excited at the first resonant mode, and 10 samples when the aluminum plate excited at the second resonant mode). When the simulation and experimental data were compared, inaccuracies were caused by nonideal structural stability and material attributes, as well as reinforcement of the electrochemical reactor.

The morphological parameters (D_p, D_c, and P) of the nanoporous AAO membranes are shown in Table 6.1.

According to SEM measurements, when no frequency excitation was applied during the anodization process, the obtained nanoporous AAO membrane had a pore diameter of 57.0 ± 10 nm, interpore distance of 120.0 ± 20 nm, and 20% porosity. When the aluminium plate reaches the first resonant mode (frequency excitation at 3.4 kHz), the obtained nanoporous AAO membrane had a pore diameter of 84.5 ± 10 nm, an interpore distance of 121.7 ± 20 nm, and a porosity of 44%. When the aluminum plate reaches the second resonant mode (frequency excitation at 4.3 kHz), the obtained AAO membrane had a pore diameter of 88.2 ± 10 nm, interpore distance of 120.9 ± 20 nm, and 48% porosity. In summary, the study showed that using resonance frequency excitation, the pore diameter increases. Consequently, porosity also increases. Nevertheless, the distance between

Table 6.1 Morphological parameters (D_p, D_c, and P) of the nanoporous AAO membrane at 3.4 and 4.3 kHz frequency excitation

Parameter	Pore diameter (nm)	Interpore distance (nm)	Porosity (%)
No excitation	57.0 ± 10	120.0 ± 20	20
First vibration mode (frequency excitation at 3.4 kHz)	84.5 ± 10	121.7 ± 20	44
Second vibration mode (frequency excitation at 4.3 kHz)	88.2 ± 10	120.9 ± 20	48

Table 6.2 Chemical composition of the AAO nanoporous membrane at frequency excitation of 3.4 and 4.3 kHz

	Element				
		Aluminum	Oxygen	Carbon	Sulfur
No excitation	Atomic concentration, at.%	35.52	62.82	1.30	0.36
	Error, %	2.3	6.3	0.2	0.1
Excitation frequency 3.4 kHz	Atomic concentration, at.%	31.65	65.78	2.18	0.39
	Error, %	2.3	7.9	0.5	0.1
Excitation frequency 4.3 kHz	Atomic concentration, at.%	35.46	63.02	1.21	0.31
	Error, %	2.2	7.1	0.3	0.1

the pores has been shown to be frequency independent. Accordingly, there was no effect on the thickness of the membrane, since all membranes had a thickness of 44 ± 5 μm.

Energy dispersive X-ray spectroscopy (EDS) analysis allowed qualitative and quantitative determination of the surface chemical composition of the nanoporous AAO membrane. The surface chemical composition of the nanoporous AAO membrane is shown in Table 6.2.

The results of the study showed that Al_2O_3 predominates. This indicated successful oxide production using high-frequency excitation. In addition to aluminum and oxygen, small amounts of carbon and sulphur, which were impurities due to the amorphous AAO, were also detected. No significant changes in the chemical composition were observed when comparing the membranes produced under different conditions of high-frequency excitation. Therefore, it was concluded that the elemental composition of the surface of the porous AAO membranes does not depend on the high-frequency excitation during the two-step anodization process.

The effect of vibrations on the pore diameter of the nanoporous AAO membrane was analysed based on the basis of theoretical growth models of porous AAO. Due to the high-frequency vibrations of the aluminum plates, the electrolyte inside the reactor was better mixed at the interface between the electrolyte and the oxide. As a result of the created fluid flow, the electrolyte was constantly renewed at the oxide and electrolyte interface. Ensuring a better mixing process of the electrolyte, the temperature and concentration of the electrolyte were constant. Since during conventional anodization, the concentration and temperature of the electrolyte changes, high-frequency oscillations have been used to obtain high-efficiency oxide growth. This led to changes in the pore diameter of the nanoporous AAO membrane.

Using a well-described two-step anodization process under high-frequency excitation conditions, the results confirm that the resonant frequency affects the geometrical parameters of the AAO nanopore structure. Such confirmatory results are useful for synthesizing and improving the structure as well as the quality of nanoporous AAO membranes.

6.2 The Effect of Ultrasound on the Formation of the AAO Nanopore Structure

The electrochemical reactor with the vibrating functional element was used to create nanoporous AAO membranes. This included vibrating an aluminum plate during the two-step anodization process. The principle of the production process is presented in Sect. 5.3. The findings indicate that the application of high-frequency excitation in the two-step anodization process allows for the control of membrane pore thickness by adjusting the ultrasonic frequency. To determine the influence of the ultrasonic frequency on the formation of the AAO nanopore structure, the aluminum plate was vibrated in the resonant frequency of the ultrasonic piezoelectric ceramic transducer. The resonant frequency of the ultrasonic piezoelectric ceramic transducer was 40 kHz. For data comparison, a frequency of 20 kHz was also investigated. Throughout the anodization process, a lightweight accelerometer was affixed to oscillating aluminum plates to observe vibration modes.

After the nanoporous AAO membranes were produced, the findings were examined using *ImageJ* software (*National Institutes of Health*, JAV) to process the SEM images of the nanoporous AAO membranes. During the experiment, 30 nanoporous AAO membranes were synthesised (10 samples without vibration, 10 samples when the 20 kHz frequency was used, and 10 samples when the 40 kHz frequency was used). The morphological parameters (D_p, D_c, and P) of the nanoporous AAO membranes are shown in Table 6.3.

During the two-step anodization process, when no frequency excitation was applied the obtained nanoporous AAO membrane had a pore diameter of 105 ± 10 nm, the interpore distance of 140 ± 20 nm, and 51% porosity. When 20 kHz frequency excitation was applied, the obtained nanoporous AAO membrane had the pore diameter of 104 ± 10 nm, interpore distance of 143 ± 20 nm, and 48% porosity. When 40 kHz frequency excitation was applied, obtained AAO membrane had the pore diameter of 105 ± 10 nm, the interpore distance of 142 ± 20 nm, and 50% porosity. Summarizing, the research showed that ultrasonic frequency excitation had no effect on pore diameter. Accordingly, there was no effect on porosity. Likewise, ultrasonic frequency excitation had no effect on the interpore distance. However, changes in membrane thickness were observed during the fabrication of nanoporous AAO membranes using ultrasonic frequencies. A microscope (*Nikon Eclipse lv150*, Tokyo, Japan) was used to measure the thickness of the nanoporous

Table 6.3 Morphological parameters (D_p, D_c, and P) of the nanoporous AAO membrane at 20 and 40 kHz frequency excitation

Parameter	Pore diameter (nm)	Interpore distance (nm)	Porosity (%)
No excitation	105 ± 10	140 ± 20	51
20 kHz frequency excitation	104 ± 10	143 ± 20	48
40 kHz frequency excitation	105 ± 10	142 ± 20	50

Table 6.4 Chemical composition of AAO nanoporous membranes at 20 and 40 kHz frequency excitation

	Element				
		Aluminum	Oxygen	Carbon	Sulfur
No excitation	Atomic concentration, at.%	35.52	62.82	1.30	0.36
	Error, %	2.3	6.3	0.2	0.1
Excitation frequency 20 kHz	Atomic concentration, at.%	31.98	64.97	2.65	0.40
	Error, %	2.2	6.8	0.4	0.1
Excitation frequency 40 kHz	Atomic concentration, at.%	37.20	61.18	1.28	0.34
	Error, %	2.2	6.1	0.2	0.1

membrane. When no vibrations were used, the thickness of the nanoporous AAO membrane was 44 μm. At 20 kHz, the thickness of the AAO membrane increased to 56 μm. At the resonant frequency of 40 kHz, the thickness of the AAO membrane also increased to 61 μm. Later, the surface chemical composition of the nanoporous AAO membrane was determined. The chemical compositions of the nanoporous AAO membranes are shown in Table 6.4.

Research showed that aluminum and oxygen predominated. This indicated a successful oxide production using ultrasonic frequency excitation. The chemical composition of the membranes produced under various conditions of high-frequency excitation was not significantly altered. Hence, it was deduced that the elemental composition of the surface of porous AAO membranes is independent of the high-frequency excitation during the two-step anodization process.

The effect of ultrasonic frequency excitation on the thickness of the nanoporous AAO membrane was analysed based on theoretical models of porous AAO growth. For this purpose, additional theoretical simulations were performed. It was analysed how high-frequency vibrations affected the fluid flow mixing process inside the membrane nanopores [9]. *COMSOL Multiphysics 6.0* software (*COMSOL*®, Burlington, United States) was used to model the fluid flow inside nanopores.

The two nanopores and the electrolyte container included the numerical model. To investigate the actual flow of the electrolyte flux between the pores as opposed to the flow demonstrated by the single pore, two pores were used. According to the experimental results, the diameter of the pore was 105 nm. The container of electrolyte represented part of the reactor volume. Finite tetrahedron elements with fixed support constraint boundary conditions were used to mesh the model. The model consisted of 3496 elements with 360 boundary elements. The walls of the pores were solid, non-slip, and subjected to isothermal boundary conditions.

The numerical model's boundary conditions as well as the computational mesh for the pores are presented in Fig. 6.1.

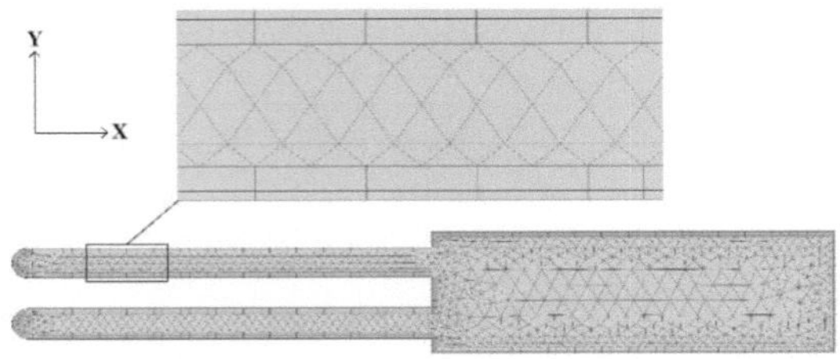

Fig. 6.1 Geometry and computational mesh of AAO nanopores

In the acoustic model, the temperature was 5 °C and the equilibrium pressure was 1 atm. The following are the parameters of the simulation model of fluid flow in nanopores: frequency 2, 20, 40 kHz; angular frequency 12,566, 1.257×10^5, 2.513×10^5 Hz; speed of sound in water 1495.3 m/s; wavelength 0.747650 m; wave number 8.4039, 84.039, 168.08 1/m; channel cross section width 105 nm; channel cross section height 1000, 5000, 45,000, 55,000 nm; wall displacement 1, 5 nm.

In the model, the velocity boundary conditions were set to be different in the x and y directions. Movements in the x-direction were unconstrained (periodic oscillation was chosen) while those in the y-direction were constrained.

The domain was chosen as a Thermoviscous Acoustics Model and was continuous. Two sets of governing equations were applied to get the outcomes. The Thermoviscous Acoustics Module was used to first construct the acoustic velocity field in the frequency domain based on the temporal and spatial scales. Next, the creeping flow module was used to calculate the streaming flow velocity field.

The computation of mean values was utilized to analyse variations in fluid flow at various excitation frequency values and pore depths as well as to analyse large amounts of data. As a result, the average values of acoustic pressure, acoustic velocity, and velocity magnitude were shown to depend on the pore depth and frequency variation. At 1 and 5 μm depths, data were recorded in 5 nm steps, and at 45 and 55 μm depths, data were recorded in 10 nm steps. Various parameters' relationships to depth and frequency are presented in Fig. 6.2.

From the curves obtained, the acoustic pressure decreases as the pore gets deeper. High-frequency excitation of 40 kHz produced the highest values of acoustic pressure. The findings of the analysis of the acoustic velocity indicated that the velocity increased marginally with the depth of the pores. Furthermore, the acoustic velocity reached its highest value at 40 kHz. The highest average fluid flow velocity values were obtained similarly at the 40 kHz frequency when analysing the velocity magnitude. The simulation results of the fluid flow velocity contour at a frequency excitation of 40 kHz at different pore depths are presented in Fig. 6.3.

At the resonant frequency of 40 kHz, the highest values were recorded. Therefore, the variation of the velocity value in the pore at different pore depths at a frequency of 40 kHz was analysed. The comparative velocity curves are presented in Fig. 6.4.

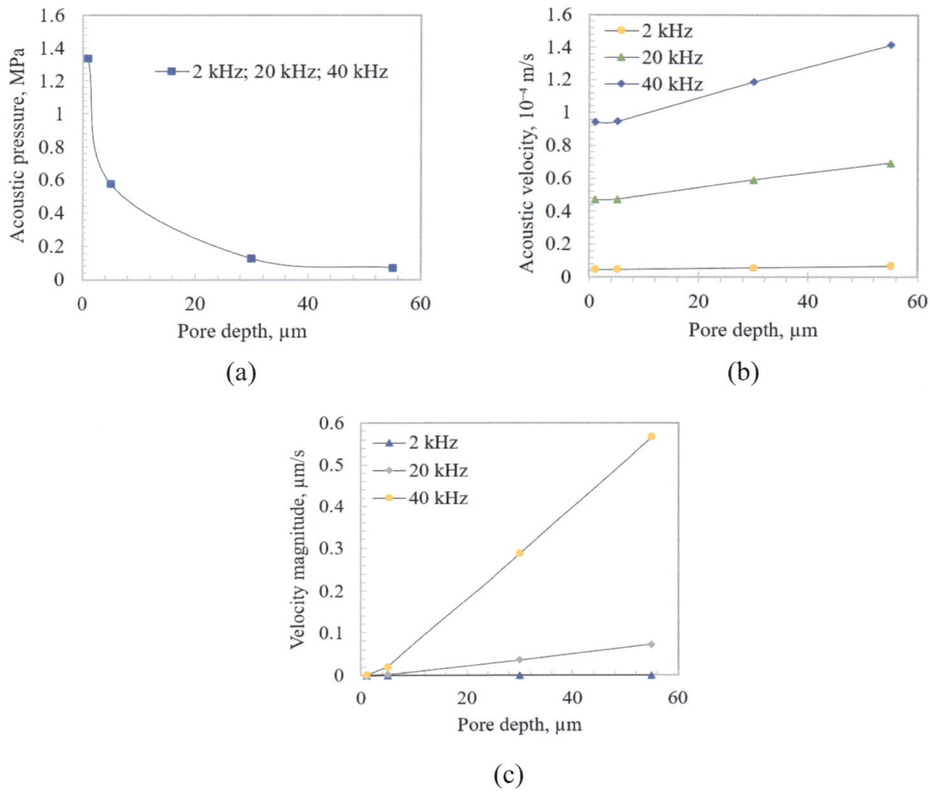

Fig. 6.2 Comparative curves of different frequencies: **a** acoustic pressure; **b** acoustic velocity; **c** velocity magnitude

The results showed that the velocity in the pores decreases uniformly. Despite the results of a 1 μm depth pore, barrier boundaries were formed at all other depths where the fluid velocity was zero. Due to these zones, by slightly changing the vibration frequency, it was possible to regulate the flow of the liquid inside the pore. This resulted in a more uniform electrolyte temperature and concentration inside the pore of the AAO membrane.

When ultrasonic frequency was used in the two-step anodization process, the electrolyte flowed, and the mixing process occurred along the entire length of the pore due to vibrations. The pores of the membrane deepen when aluminum oxide is formed. In the deeper pore, the direction of the liquid flow had changed. In addition, a liquid flow barrier appeared where the liquid mixing process did not take place. The electrolyte flow rate was highest at the beginning of the pore, where the electrolyte entered the pore from the reactor. The electrolyte near the pore was intensively mixed with the entire reactor liquid, and the flow of liquid entering the pore was constantly renewed. The numerical

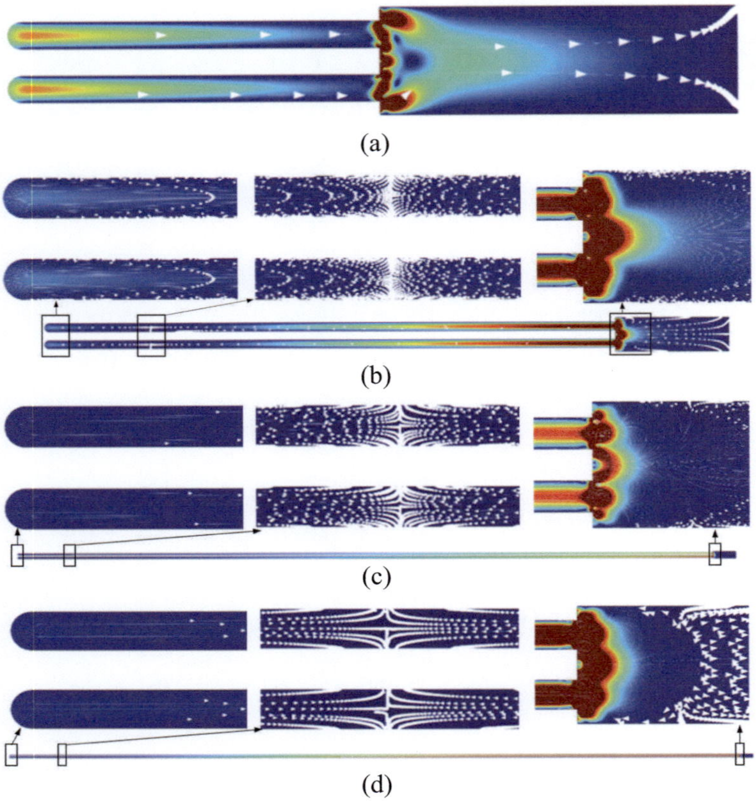

Fig. 6.3 Fluid flow velocity contour at a frequency excitation of 40 kHz at different pore depths: **a** 1 μm; **b** 5 μm; **c** 45 μm; **d** 55 μm

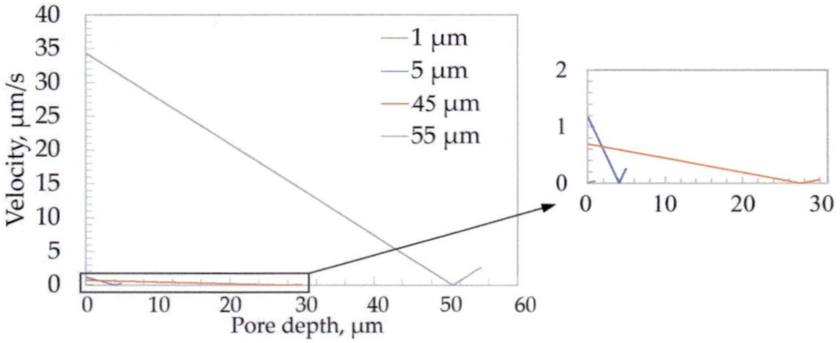

Fig. 6.4 Comparative velocity curves of different pore depths

simulation results confirmed that the mixing of the liquid flow in the pores was ensured by high-frequency excitation using the ultrasonic frequency. Based on theoretical models of oxide growth, it was concluded that high-frequency excitation during the anodization process ensured a more uniform distribution of electrolyte temperature and pH values along the entire pore length. Therefore, during mixing, the concentration and temperature of the electrolyte in the pores were updated, which led to faster oxide formation. This explains the increased thickness of the aluminum oxide membrane using the ultrasonic frequency during the two-step anodization process.

References

1. Osama L, Handal HT, El-Sayed SAM, Elzayat EM, Mabrouk M (2024) Fabrication and optimisation of alumina nanoporous membranes for drug delivery applications: a comparative study. Nanomaterials 14(13):1078. https://doi.org/10.3390/nano14131078
2. Liu S, Tian J, Zhang W (2021) Fabrication and application of nanoporous anodic aluminum oxide: a review. Nanotechnology 32:222001. https://doi.org/10.1088/1361-6528/abe25f
3. Peng Q, Xie X, Xu Q, Lan T, Sun C, Zhang L, Dong M (2020) The effect of the current pulse amplitude on the nanopore structures of 3D-AAO films. Microporous Mesoporous Mater 309:110575. https://doi.org/10.1016/j.micromeso.2020.110575
4. Ayalew AA, Han X, Sakairi M (2024) Effect of substrate temperature and electrolyte composition on the fabrication of through-hole porous AAO membrane with SF-MDC. Mater Chem Phys 323:129658. https://doi.org/10.1016/j.matchemphys.2024.129658
5. Chernyakova K, Tzaneva B, Vrublevsky I, Videkov V (2020) Effect of aluminum anode temperature on growth rate and structure of nanoporous anodic alumina. J Electrochem Soc 167:103506. https://doi.org/10.1149/1945-7111/ab9d65
6. Eessaa AK, El-Shamy AM (2023) Review on fabrication, characterization, and applications of porous anodic aluminum oxide films with tunable pore sizes for emerging technologies. Microelectron Eng 279:112061. https://doi.org/10.1016/j.mee.2023.112061
7. Sener M, Sisman O, Kilinc N (2023) AAO-assisted nanoporous platinum films for hydrogen sensor application. Catalysts 13(3):459. https://doi.org/10.3390/catal13030459
8. Laucirica G, Terrones YT, Toimil-Molares ME, Trautmann C, Marmisollé WA, Azzaroni O (2024) Membrane nanoarchitectonics: advanced nanoporous membranes for osmotic power generation. In: Materials nanoarchitectonics from integrated molecular systems to advanced devices micro and nano technologies, chap 3, pp 29–46. https://doi.org/10.1016/B978-0-323-99472-9.00021-3
9. Cigane U, Palevicius A, Janusas G (2022) Vibration-assisted synthesis of nanoporous anodic aluminum oxide (AAO) membranes. Micromachines 13:2236. https://doi.org/10.3390/mi13122236

A Free-Standing Chitosan Membrane Prepared by the Vibration-Assisted Solvent Casting Method

7

Introduction. Chitosan is a biopolymer that has applications in various fields due to its specific properties, including biocompatibility, biodegradability, low toxicity, immunogenicity, and antibacterial activity. In recent years, there has been high interest in the potential applications of chitosan membranes. Research indicates that chitosan exhibits great potential as a material for the creation of skin barriers that are bactericidal, biocompatible, and bioadhesive. It should be noted that commercial wound dressings [1–4] and hemostatic dressings [5] already frequently contain chitosan. So, surface modification of artificial skin barriers for the treatment of skin tissue damage has received a lot of attention. To improve the surface properties of chitosan membranes, various surface modifications have been performed. A promising way to control the surface area of chitosan membranes is to create a nanopillar surface. In fact, in biological applications, including tissue engineering and medical treatment, nanopillar surfaces are more advantageous than conventional planar surfaces due to their unique features [6–9]. The three impacts of nanopillars on surface biological processes are as follows: they enhance the surface area, improve the penetration capacity of the cell, and promote cell adhesion [10]. The use of templates is one technique to create nanopillars. When the template with the required structure is available, a chitosan solution is poured onto the template. This membrane production method is simple and is known as the solvent casting method. However, it becomes challenging to guarantee that the solution flows into the nanopores using the solvent casting method when the AAO template bottom holes are closed. Since chitosan membrane production technologies that involve membrane structural geometry control are relevant, this chapter introduces the vibration-assisted solvent casting method. Nanopillared chitosan membranes were produced using nanoporous AAO membranes as template. The high-frequency oscillations were used to ensure and control the flow of the chitosan

A. Palevicius et al., *Nano/Micro Functional Elements Formation for Bioengineering Applications*, Synthesis Lectures on Biomedical Engineering,
https://doi.org/10.1007/978-3-031-81509-6_7

solution into the pores of the AAO membrane. In this way, the height of the nanopillars formed on the surface of the chitosan membrane could be controlled by high-frequency vibrations.

7.1 Fabrication of Chitosan Membrane

Nanoporous AAO templates were used for the development of nanopillared chitosan membranes. Nanoporous AAO templates have been produced using the two-step anodization technique, which is described in Sect. 5.1. At a temperature of 5 °C and a voltage of 60 V, the anodization process was started. During the first hour-long anodization procedure, high-frequency was not applied. The second step of the anodization process was conducted for eight hours at 60 V and 5 °C in the electrochemical reactor. High-frequency excitation of 40 kHz had been used in the anodization process during the second stage. Finally, the produced nanoporous AAO membranes were cleaned using distilled water, allowed to dry in air, and then used as templates to create nanopillared chitosan membranes. Figure 7.1 illustrates how the nanopillared chitosan membrane was produced.

To produce chitosan membranes, high molecular weight chitosan with a degree of deacetylation >95% was used. To dissolve chitosan, 1 percent acetic acid was used. Chitosan in an acetic acid solution was stirred at 20 °C for 1 h until the chitosan was completely dissolved. In the next step, the chitosan solution was dropped onto the prepared nanoporous AAO templates and dried at 20 °C for 24 h. The casting temperature was 20 °C. The diameter of the nanoporous AAO templates was 40 mm. After the complete drying of the chitosan membrane, the template was removed. For that purpose, the samples were immersed in 0.5 mol/l NaOH solution. Finally, the nanopillar chitosan membranes were ultrasonically cleaned in distilled water. Using the same production principle, 10 chitosan membranes were produced. The surface geometry of all chitosan membranes was investigated using SEM.

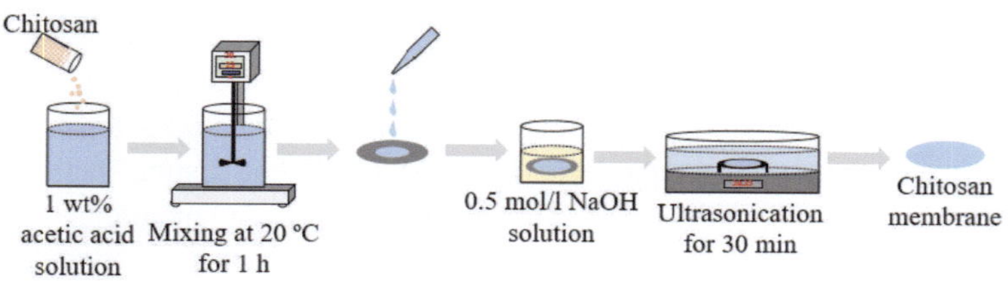

Fig. 7.1 Schematic representation of fabrication procedure of the chitosan membrane

7.2 Improved Method of Solvent Casting Using High-Frequency Vibrations

To determine the influence of vibrations on the geometry of the nanopillars and to fabricate nanopillars of different heights on the surface of the chitosan membrane, an improved method of solvent casting using high-frequency vibrations was used [11]. The fabrication of nanopillared chitosan membranes using high-frequency excitation is presented in Fig. 7.2.

The detailed device design for chitosan and AAO template vibrations is shown in Fig. 7.3.

All steps of the standard solvent casting method are described in detail in Sect. 7.1. An improved method of solvent casting used high-frequency vibrations. During the study, three chitosan solutions of different concentrations were prepared. After dropping the chitosan solution onto the nanoporous AAO template, the chitosan solution was vibrated together with the AAO template at a frequency of 35 kHz for 5 s. Prepared 3 chitosan solutions of different concentrations using 1, 2 and 3 wt% chitosan solution in 1 wt% acetic acid solution. Using the same production principle, for each concentration, 10 membranes were produced. The surface geometry of all nanopillared chitosan membranes was investigated using SEM.

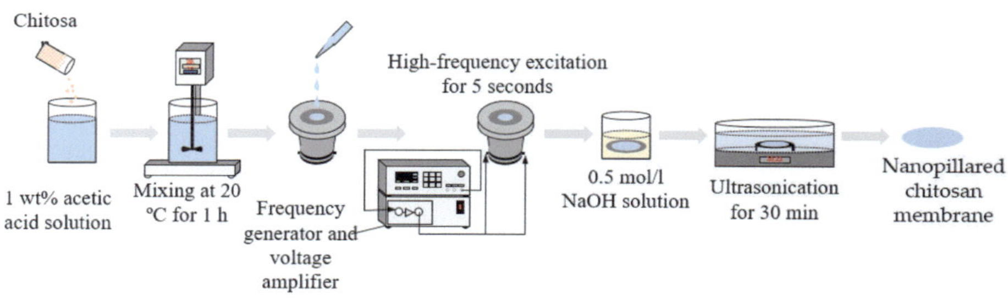

Fig. 7.2 Fabrication of a nanopillared chitosan membrane using high-frequency excitation

Fig. 7.3 Device for the high-frequency excitation of chitosan solution and nanoporous AAO template

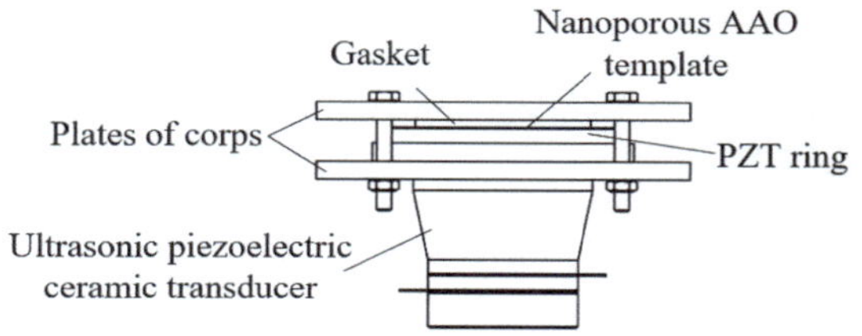

7.3 The Effect of the High-Frequency on the Formation of Nanopillars on Chitosan Membrane Surface

Using the vibration-assisted solvent casting method, 30 free-standing chitosan membranes (ten membranes for each concentration) were fabricated. To determine the influence of 35 kHz frequency excitation during the solvent casting method on the height of the nanopillars, chitosan solutions of three different concentrations were used. The nanopillared chitosan membranes were experimentally prepared and details of their fabrication are described in Sect. 7.2. For the study, chitosan solutions of 1, 2, and 3 wt% concentration were used. The SEM images of the chitosan membranes were analysed using *ImageJ* software (*National Institutes of Health*, JAV). The results are presented in Table 7.1.

When vibrations were not used during the solvent casting method, nanopillars were not formed on the surface of the chitosan membrane. On the other hand, nanopillared chitosan membranes were produced when 35 kHz frequency excitation was used during the solvent casting method. Depending on the concentration of the prepared chitosan solution, nanopillars of different heights were obtained. Furthermore, since the diameters of the nanopillars were 100 ± 10 nm, it was observed that the diameter of the nanopillars was independent of the concentration.

The unique surface structure of the nanopillared chitosan membrane is important for its application in different areas. The initial surface area is greatly increased by a large number of vertically aligned nanopillars, without affecting the original substrate's size. The surface area (S) of the nanopillared surface could be calculated using formula [12]:

$$S = S_0 + n(2\pi rl) \tag{7.1}$$

where S_0—the original flat surface area; r—the nanopillar radius; l—the nanopillar height; n—the number of nanopillars on the flat surface.

To create unique artificial skin barriers for chitosan, it is important to ensure a larger surface area of the chitosan membrane. For this purpose, the calculation of the surface area of 20×20 mm freestanding chitosan membrane sheet was calculated. The free-standing chitosan membrane is presented in Fig. 7.4.

The experimental results of nanopillared chitosan membranes fabricated of different concentrations are presented in Table 7.2.

Table 7.1 Experimental results of a nanopillared fabricated chitosan membrane

Parameter	1 wt%	2 wt%	3 wt%
Diameter of nanopillars, nm	101 ± 10	102 ± 10	100 ± 10
Height of nanopillars, nm	1001 ± 10	666 ± 10	387 ± 10
Average surface area of a single nanopillar, μm^2	0.325	0.221	0.129

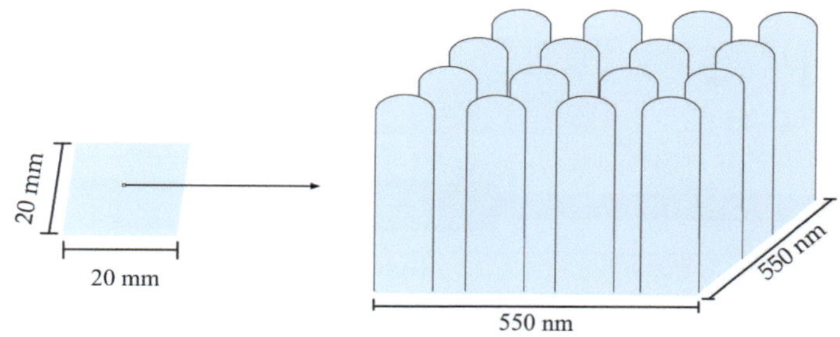

Fig. 7.4 Chitosan membrane for artificial skin barriers' creation

Table 7.2 Surface areas of nanopillared chitosan membranes

Parameter	1 wt%	2 wt%	3 wt%
Diameter of nanopillars, nm	101 ± 10	102 ± 10	100 ± 10
Height of nanopillars, nm	1001 ± 10	666 ± 10	387 ± 10
Average surface area of a single nanopillar, μm^2	0.325	0.221	0.129
The number of nanopillars	1.6×10^{11}		
Calculated surface area, cm^2	56.00	39.36	24.64

When the chitosan solution of different concentrations was poured onto the nanoporous AAO template using 35 kHz excitation for 5 s, the experimental results of the surface area calculation showed that larger surface areas of the free-standing chitosan membrane were obtained.

Furthermore, the experimental results demonstrate that different heights of the nanopillars produced using vibrations inspire theoretical calculations based on the acoustics and concentration of the chitosan solution.

Acoustics can be used to control microparticles, living cells, and other particles because of the recent progress in the development of microfluidic devices [13]. Particles are impacted by a viscous drag force as well as an acoustic streaming flow generated in the microchannels using acoustics [14]. The particle trajectories are controlled by the forces of acoustic radiation and viscous drag [15]. The effect of the transfer of momentum from the acoustic field to the particles is the force of acoustic radiation, which is caused by non-linear factors in the governing equations [16]. Consequently, particles are affected by a net force of the acoustic radiation force [17].

For research, the multiphysics task, which had the three main phases, was used as a theoretical model. The following are the three main phases: (1) the first-order acoustic field in the domain was identified; (2) the acoustic streaming flow in the domain has been added; (3) the fluid flow inside the pore was evaluated along with the streaming flow.

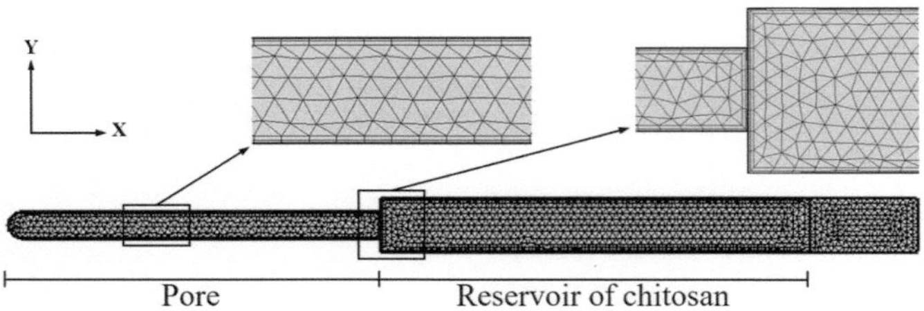

Fig. 7.5 The finite element mesh of the simulation model of the flow of the chitosan solution into the pore

In this work, fluid behaviour in the nanopore was simulated using the *COMSOL Multiphysics 6.0* software (*COMSOL*®, Burlington, United States). High frequencies were used to model the deposition of the chitosan solution on the porous AAO structure and the liquid's access into the pores. The mathematical model consisted of a chitosan reservoir and a single pore with 100 nm diameter. The model was meshed by free triangular elements. In the model, the boundary layer was inside the structure. The finite element mesh is shown in Fig. 7.5.

The following are the parameters of the simulation model of chitosan solution flow in nanopores: frequency 35 kHz, ambient temperature 20 °C, ambient pressure 1 atm, study angular frequency 2.5132×10^5 Hz, channel cross-section width 100 nm, channel cross-section height 1000 nm, wall displacement 100 nm, time 5 s. The model consisted of 5718 domain elements and 494 boundary elements. Periodic movements took place in the x direction, while movements in the y direction were restrained. The inner wall was a solid surface with no-slip and isothermal boundary conditions.

The model was subjected to level set physics, laminar flow, and thermoviscous acoustics. The two-phase flow was chosen for the multiphysics discipline. In this case, the fluid flow was selected as the laminar flow.

After performing theoretical calculations, when no high-frequency excitation was used, the flow of the chitosan solution into the closed nanopore was not ensured. The simulation results are presented in Table 7.3.

When high frequency excitation of 35 kHz in 5 s was applied, the velocity of the 1 wt% chitosan solution was 245 nm/s. Under the same settings, the velocity of the

Table 7.3 Simulation results of the flow of the chitosan solution into the pore

Parameter	1 wt%	2 wt%	3 wt%
Velocity, nm/s	245	155	88
Height of nanopillars, nm	1225	775	440
Surface area of a single nanopillar, μm^2	0.393	0.251	0.146

2 wt% chitosan solution was 155 nm/s and the velocity of the 3 wt% chitosan solution was 88 nm/s.

The simulation results confirmed the propagation of sound waves. Waves travel through the air in pores. Also, the waves propagated through the chitosan solution above the pore. All this was shown by the acoustic velocity. Affected by the streaming velocity, the chitosan solution gained energy and could flow into the pores. Additionally, the flows of three different concentrations of chitosan solutions into the pores were computed. The final simulation pictures of AAO pores and chitosan solution are shown in Fig. 7.6.

For comparison of theoretical and experimental results, the experimental velocity of the chitosan solution in the pores of AAO membranes was calculated. The comparison is presented in Table 7.4.

Theoretical and experimental results are slightly different due to non-ideal conditions. The obtained results confirm that the heights of the nanopillars can be controlled by using the prescribed high-frequency vibration and chitosan solutions with different concentrations. The different heights of the nanopillars make it possible to ensure a certain surface area.

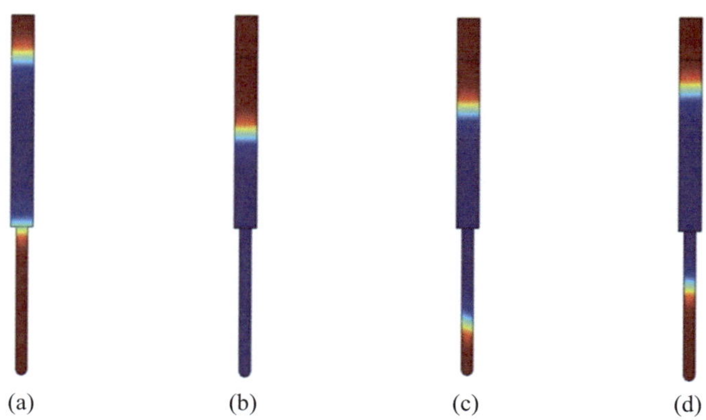

Fig. 7.6 Flow of different concentrations of chitosan solution into the nanopore using the high-frequency excitation of 35 kHz in 5 s (blue represents the chitosan solution and red represents the air): **a** any concentration of chitosan solution without high-frequency excitation; **b** 1 wt% chitosan solution; **c** 2 wt% chitosan solution; **d** 3 wt% chitosan solution

Table 7.4 Comparison of the theoretical and experimental velocity of the chitosan solution in a nanopore

Parameter	1 wt%	2 wt%	3 wt%
Theoretical velocity, nm/s	245	155	88
Calculated experimental velocity, nm/s	200	133	77

Due to the easily controlled surface area using high-frequency oscillations for 5 s during the solvent casting method and the use of chitosan solutions of different concentrations, these studies contribute to the development of nanostructured membranes. Moreover, the results could be useful in the development of artificial skin barriers for commercial use.

References

1. Lin HT, Venault A, Chang Y (2019) Zwitterionized chitosan based soft membranes for diabetic wound healing. J Membr Sci 591:117319. https://doi.org/10.1016/j.memsci.2019.117319
2. Augustine R, Rehman SRU, Ahmed R, Zahid AA, Sharifi M, Falahati M, Hasan A (2020) Electrospun chitosan membranes containing bioactive and therapeutic agents for enhanced wound healing. Int J Biol Macromol 156:153–170. https://doi.org/10.1016/j.ijbiomac.2020.03.207
3. Genesi BP, Barbosa RM, Severino P, Rodas ACD, Yoshida CMP, Mathor MB, Lopes PS, Viseras C, Souto EB, Silva CF (2023) Aloe vera and copaiba oleoresin-loaded chitosan films for wound dressings: microbial permeation, cytotoxicity, and in vivo proof of concept. Int J Pharm 634:122648. https://doi.org/10.1016/j.ijpharm.2023.122648
4. Ji M, Li J, Wang Y, Li F, Man J, Li J, Zhang C, Peng S, Wang S (2022) Advances in chitosan-based wound dressings: modifications, fabrications, applications and prospects. Carbohyd Polym 297:120058. https://doi.org/10.1016/j.carbpol.2022.120058
5. Hamedi H, Moradi S, Hudson SM, Tonelli AE, King MW (2022) Chitosan based bioadhesives for biomedical applications: a review. Carbohyd Polym 282:119100. https://doi.org/10.1016/j.carbpol.2022.119100
6. Han W, Ren J, Xuan H, Ge L (2018) Controllable degradation rates, antibacterial, free-standing and highly transparent films based on polylactic acid and chitosan. Colloids Surf A 541:128–136. https://doi.org/10.1016/j.colsurfa.2018.01.022
7. Jiang Y, Deng Y, Tu Y, Ay B, Sun X, Li Y, Wang X, Chen X, Zhang L (2019) Chitosan-based asymmetric topological membranes with cell-like features for healthcare applications. J Mater Chem B 7:2634–2642. https://doi.org/10.1039/C8TB03296C
8. Jenkins J, Mantell J, Neal C, Gholinia A, Verkade P, Nobbs AH, Su B (2020) Antibacterial effects of nanopillar surfaces are mediated by cell impedance, penetration and induction of oxidative stress. Nat Commun 11:1626. https://doi.org/10.1038/s41467-020-15471-x
9. Valiei A, Lin N, McKay G, Nguyen D, Moraes C, Hill RJ, Tufenkji N (2022) Surface wettability is a key feature in the mechano-bactericidal activity of nanopillars. ACS Appl Mater Interfaces 14:27564–27574. https://doi.org/10.1021/acsami.2c03258
10. Gudur A, Ji HF (2016) Bio-applications of nanopillars. Front Nanosci Nanotechnol 2:1–10. https://doi.org/10.15761/FNN.1000140
11. Cigane U, Palevicius A, Janusas G (2023) A free-standing chitosan membrane prepared by the vibration-assisted solvent casting method. Micromachines 14:1419. https://doi.org/10.3390/mi14071419
12. Ngo TD, Kashani A, Imbalzano G, Nguyen KTQ, Hui D (2018) Additive manufacturing (3D printing): a review of materials, methods, applications and challenges. Compos B Eng 143:172–196. https://doi.org/10.1016/j.compositesb.2018.02.012
13. Lei J, Cheng F, Li K (2020) Numerical simulation of boundary-driven acoustic streaming in microfluidic channels with circular cross-sections. Micromachines 11:240. https://doi.org/10.3390/mi11030240

14. Peng T, Zhou M, Yuan S, Fan C, Jiang B (2022) Numerical investigation of particle deflection in tilted-angle standing surface acoustic wave microfluidic devices. Appl Math Model 101:517–532. https://doi.org/10.1016/j.apm.2021.07.018
15. Qiao Y, Zhang X, Gong M, Wang H, Liu X (2020) Acoustic radiation force and motion of a free cylinder in a viscous fluid with a boundary defined by a plane wave incident at an arbitrary angle. J Appl Phys 128:044902. https://doi.org/10.1063/5.0005866
16. Wang Q, Riaud A, Zhou J, Gong Z, Baudoin M (2021) Acoustic radiation force on small spheres due to transient acoustic fields. Phys Rev Appl 15:044034. https://doi.org/10.1103/PhysRevApplied.15.044034
17. Basch T, Pavlic A, Dual J (2019) Acoustic radiation force acting on a heavy particle in a standing wave can be dominated by the acoustic microstreaming. Phys Rev E 100:061102. https://doi.org/10.1103/PhysRevE.100.061102

Application for Bioengineering 8

8.1 Plasmon Metal Nanostructures for Detection and Sensing Cell-Biological Particles

This subchapter proposes a novel piezoelectric composite material whose basic element is PZT and a sensing platform where this material was integrated. Results showed that a designed novel cantilever-type element is able to generate a voltage up to 80 μV at 50 Hz frequency. To use this element for sensing purposes, a four microns periodical microstructure was imprinted. Silver nanoparticles were precipitated on the grating to increase the sensitivity of the designed element, i.e. Surface Plasmon Resonance (SPR) effect appears in the element. To tackle some issues (a lack of sensitivity, signal delays) the element must have certain electronic and optical properties. And a combination of piezoelectricity and SPR in a single element is one possible solution, proposed in this subchapter [1].

Introduction. Past few decades biomedical applications have required fast, reliable, miniature and low-cost methods and tools for recognition of bio-molecules in various fluids. One of the recent new applications in this area is related to bio-sensing elements based on cantilever type sensing platforms. These platforms are able to convert biological responses into electrical signals [2–4].

This subchapter discusses a design of a cantilever-type sensing element made of a novel piezoelectric material exhibiting high resonance frequencies, leading to a faster response time and much higher sensitivity compared to cantilevers made of silicon oxide and etc. The advantage of the proposed design in this subchapter is a periodical microstructure imprinted on top of the piezoelectric layer with metal nanoparticles precipitated on the grating ridges. Because of incorporation of noble (in this case silver) nanoparticles, the SPR effect appears. This effect highly influences the efficiency, the structure and operation itself, i.e. much greater control of optical properties, sensitivity and selectivity may be

A. Palevicius et al., *Nano/Micro Functional Elements Formation for Bioengineering Applications*, Synthesis Lectures on Biomedical Engineering,
https://doi.org/10.1007/978-3-031-81509-6_8

achieved. Moreover, to achieve the maximum optical effect, an operating wavelength of the sensing element may be tuned to a spectral region where SPR peak is sharpest. SPR is a powerful tool for investigating biomolecular interactions with label-free real-time analytical technologies.

Many various materials have a property of piezoelectricity, but only few of them are most promising in MEMS and NEMS technologies, i.e. zinc oxide (ZnO) films [5], polyvinylidene fluoride (PVDF) films [6] and lead zirconate titanate (PZT) [7], so-called, polycrystalline ceramics. These three main materials have high piezoelectric coefficients, a very good flexibility and a strong electromechanical coupling. Moreover, it is known that nano-sized ceramics are very different compared to bulk ceramics in their mechanical behavior. This research was concentrated on designing a novel piezoelectric material, working at low frequencies and able to harvest energy or to cause deformations. This novel material may be integrated in sensing or actuating elements, depending on the purpose of microsystem. Economical and easy fabrication allows it to use in nowadays technologies. In this research, PZT (exhibiting high piezoelectric coefficient and permittivity, large dielectric constants and good conversion efficiency) was chosen as a basic material for creating a novel piezoelectric sensing platform. A mixture of 20% polyvinyl butyral together with PZT powder was synthesized. Since property of strong binding is essential, a polyvinyl butyral works here as a binding element with PZT ensuring good adhesion and flexibility. Three concentrations of PZT (40, 60 and 80%) composite materials were produced for profound investigations. Each PZT material was coated as a thin film and sandwiched between two cooper electrode layers. Results showed that in the mode configuration of d31, it harvested energy, i.e. at 50 Hz frequency it generated up to 80 μV. Further, a 4-μm periodical microstructure was imprinted on thin 80% PZT piezoelectric film and silver (Ag) nanoparticles deposited on it. A property of piezoelectricity allows tuning a diffraction grating and results in variation of diffraction efficiencies. It allows achieving desirable spectral region in which the designed cantilever-type sensing platform would be of a high-efficiency. To perform certain specialized sensing function, it must reliably store and convert different forms of energy, transduce signals, and respond repeatedly to external biological and chemical environments. The designed cantilever-type sensing platform can alter mechanical stress within the oscillator and its total mass when target analyte is in contact with its surface. Here, a signal transduction is achieved by employing a diffraction grating to measure the mechanical bending or the frequency spectra resulting from additional loading by the absorbed mass. Since both the resonant frequency shift and deflection are highly dependent on the position of the absorbed material (analyte) it is difficult to determine the exact amount of additional mass. A diffraction grating and incorporation of Ag nanoparticles on its surface allow adequate control of chemical surface functionality for the detection of analytes of interest, i.e. defined molecules can be absorbed from analyte onto the Ag anchors, creating a strong interaction between the functional group and the silver nanoparticles. Vibrating cantilever-based platform offers quantitative assessment of the specific mass when experimentally monitoring resonant frequency shifts. Advantages of the designed novel sensing platform include: easy fabrication, inexpensive materials and equipment required, ability to make thin (from 400 nm to 1.4 μm) films; wide application areas (from sensitive diagnostic devices in medicine, pharmacy, to microsystem devices in wireless sensors and portable electronics). Interesting fact is that few different components—resonance, piezoelectricity, diffraction efficiency and silver nanoparticles, all are combined in a single element.

Experimental section. A novel piezoelectric sensing platform was created and evaluated in the Institute of Materials Science of Kaunas University of Technology and in the Institute of Mechatronics of Kaunas University of Technology.

Material, synthesis and formation. A novel piezoelectric material was designed using PZT powder and 20% solution of polyvinyl butyral in benzyl alcohol and mixing them at defined conditions. Three material types using different PZT (40, 60 and 80%) concentration were produced. Here, screen printing technique was used to cover the base of element with PZT paste on a copper foil. Using a 325 mesh of stainless steel PZT paste was printed and dried at 100 °C for 30 min. This procedure allows controlling a smooth and equal thickness of formed thin films (in this case, thickness of all 3 elements was ~ 1.2 μm). Further, a four-micron periodical microstructure was formed on the top of piezoelectric thin film by hot-embossing technique. Ag nanoparticles were formed from a solution of 0.05 M $AgNO_3$ in deionized water and dip-coated on the periodical microstructure. The purpose of using different PZT concentrations was creation of piezoelectric material with good piezoelectric characteristics and sufficient elasticity for imprint of well-defined grating.

Cantilever type piezoelectric element structure. A novel sensing element was made of a cantilever-type. For evaluation of piezoelectric properties, three samples with different PZT (40, 60 and 80%) concentrations were designed. Principal scheme of piezoelectric element is presented in Fig. 8.1.

A cantilever-based sensing platform allows precise evaluation of piezoelectric properties. The elements were investigated in both—direct and indirect piezoelectric effects.

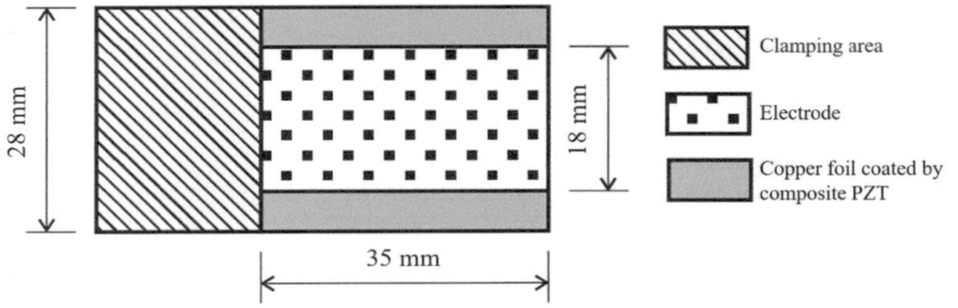

Fig. 8.1 Principal scheme of piezoelectric element

Surface morphology and chemical composition measurements. Structure and chemical composition of the designed material was investigated using scanning electron microscope (SEM) Quanta 200 FEG (Hillsboro, OR, USA). It is integrated with the energy dispersive X-Ray spectrometer (EDS) detector X-Flash 4030 from Bruker (Berlin, Germany). Samples were examined under controlled pressure water steam atmosphere. A 133 eV (at Mn K) energy resolution at 100.000 cps was achieved with a 30 mm^2 area solid state drift detector, cooled with Peltier element. X-Ray spectroscopy method allows analyzing energy distributions. The energy differences were measured between the various quantum states of a system together with the probabilities that the system jumps between these states.

Investigations of surface morphology were performed with an Atomic Force Microscope NT-206 (Micro-test machines Co) in contact mode. It is a surface analytical technique employed to generate very high-resolution topographic views of a surface down to molecular/atomic resolution, with the only requirement being that the sample be deposited on a flat surface. This method provides spatial resolutions of 1–20 nm. AFM is able to analyze electrical, magnetic, mechanical and chemical properties in nanoscale dimension, using specialized modes. Main morphology parameters: Z_{mean}—average height, R_a—arithmetic average surface roughness, R_q—root mean square surface roughness.

Harmonic excitation measurements. Harmonic excitation measurements were performed using a scheme presented in Fig. 8.2.

An experimental setup consists of a piezoelectric element, excitation measurement systems and data acquisition. An electromagnetic shaker excites the element fixed in a custom-made clamp. A harmonic excitation signal is transmitted to the electromagnetic shaker and controlled by a function generator *Agilent 33220 A* (*Agilent*, Santa Clara, CA, USA) and a voltage amplifier *HQ Power VPA2100MN* (*HQ*, Power, Gavere, Belgium). For acceleration measurements, at the top of the clamp a single-axis miniature accelerometer

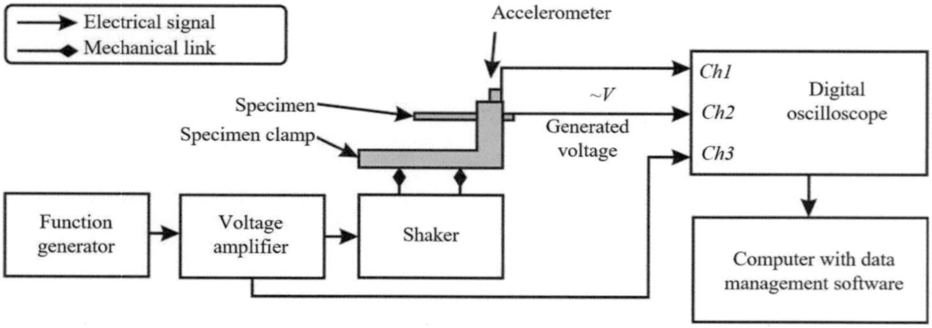

Fig. 8.2 An experimental setup scheme used to measure harmonic excitations of piezoelectric elements

KD-35 (with sensitivity of 50 mV/g ±2% and working frequency from 5 Hz to 5 kHz) (*Metra*, Radebeul, Germany) is attached. Signals from a voltage amplifier, accelerometer and an element are collected with data acquisition system, consisting of a 4-channel USB oscilloscope (analog-to-digital converter; *Pico*, St Neots, United Kingdom) *PICO 3424*, and forwarded to a computer. Obtained data is then analyzed with *Pico-Scope 6* software.

Vibration analysis. A two-beam speckle pattern interferometer, or so called PRISM system (measurement sensitivity <20 nm, measurement range >100 μm), was used to evaluate a response of electrical excitation of the investigated piezoelectric element. This method allows measuring vibration and deformation with minimal sample preparation and with no contact with the sample surface. PRISM is a rather high-speed holographic technique, equipped with a computer system and integrated software (Fig. 8.3).

The laser beam directed to the object is an object beam; the beam directed to the camera—a reference beam (Fig. 8.4).

Camera lens collects the scattered laser light from the object and images the object onto sensors of CCD camera. The reference beam falls directly to the camera and overlaps the

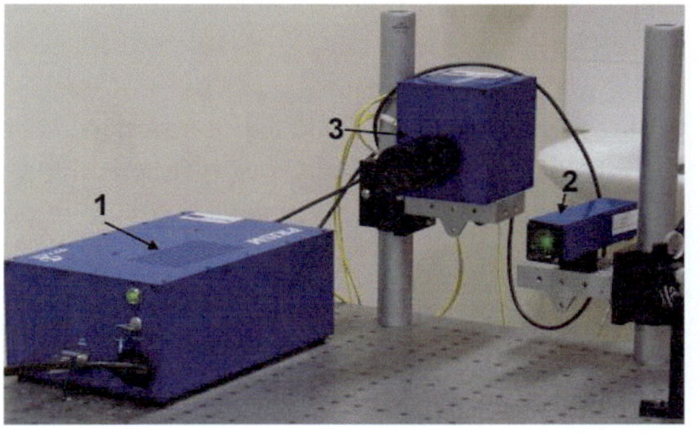

Fig. 8.3 A PRISM measurement system: 1—a control block, 2—an object illumination head, 3—a video camera

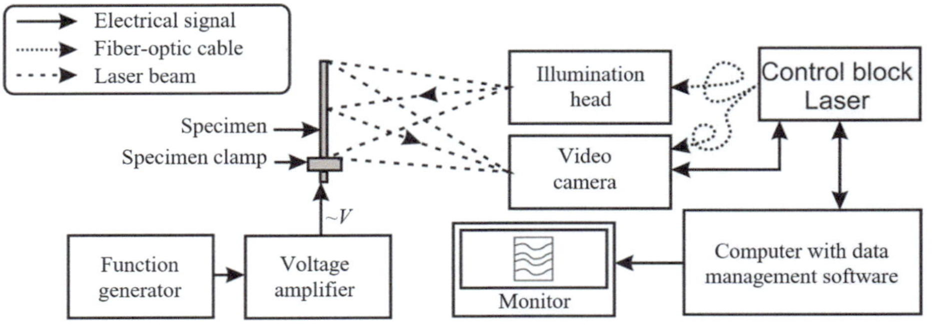

Fig. 8.4 The PRISM experimental set-up

(a) (b) (c)

Fig. 8.5 SEM view of piezoelectric elements with PZT concentration of: **a** 40%, **b** 60% and **c** 80%

image of the object. The fringes displayed on the monitor appear because of the shape changes occurring between a reference and a stressed state of the object. Obtained data allows evaluating the electrical excitations of the investigated element.

Measurements of diffraction efficiency. For evaluation of diffraction efficiency in all peaks for different incident angles, a laser diffractometer of red He–Ne laser (wavelength 632.8 nm incident to grating) was used. It consists of He–Ne laser source, set of mirrors to direct a beam, photodiode and a tester, to record the measurement data. A photodiode records the intensity and angle of diffracted light from the grating in all maxima (0, ± 1, ± 2, and etc.), which is dependent on a grating period.

Results and discussion. *Surface Morphology of Novel Cantilever Type Piezoelectric Elements.* Surface morphology and chemical properties of three different elements with PZT concentrations of 40, 60 and 80%, were investigated. Using SEM surface morphology of piezoelectric elements was investigated (Fig. 8.5).

Different sizes of grains were observed in samples: surface of the first element PZT 40% was rather smooth with small 5–15 μm diameter pileups observed on top (Fig. 8.5a). Increased PZT concentration to 60% leads to formation of individual grains (Fig. 8.5b). The surface of PZT 80% piezoelectric film became granular with smaller grain size below 4 μm (Fig. 8.5b). PZT particle loaded in a polyvinyl butyral might be the origin of the irregular shape, nucleation and growth in the solution, thus forming the smaller grains group. PZT island structures (Fig. 8.5c) were formed where the granular grains surround larger grains. It is also seen that a high density is achieved in PZT 40 and 60% except few pinholes. However, these micro cracks with the length of micron distribute uniformly in the surface. SEM surface views in scale of 5–10 μm is presented in Fig. 8.6.

The element with PZT concentration of 40% (Fig. 8.6a) has some small structures of about 10–12 nm sizes. The element with PZT 60% concentration has a small net with

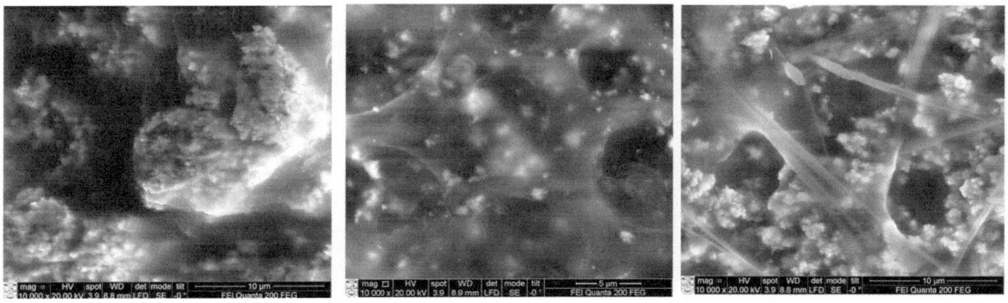

(a) (b) (c)

Fig. 8.6 Images of SEM of piezoelectric elements when PZT: **a** 40%; **b** 60%; **c** 80%

cavities formed on the surface (Fig. 8.6b). Higher PZT concentration leads to formation of three-dimensional microstructures with empty cavities of about 6–8 μm diameter (Fig. 8.6c).

Chemical composition novel of piezoelectric elements. Chemical compositions of cantilever-type piezoelectric elements were investigated with energy-dispersive (ED) spectrometer. A pulse height analysis is employed. Ionization is caused by incident X-ray photons; electrical charge is produced. Energy dispersive spectrum is displayed. The x-axis represents the X-ray energy in channels 1.5–5 eV wide and the y-axis represents the number of counts per channel up to 1600 cps/eV. Three main elements were defined in the samples: Lead (Pb), Zirconium (Zr), and Titanium (Ti). Conventionally, for the Ti K_β energy resolution peak is specified at about 4.94 keV. For Zr L_α the peak is achieved at ~2.05 eV and Pb peak is about 2.35 eV (Fig. 8.7).

The peak values of Pb, Zr and Ti of the elements (when PZT 40, 60 and 80%) for more accurate comparison are presented in Table 8.1.

Piezoelectric properties. A PRISM experimental system was used to investigate a response to electrical excitation of designed novel cantilever-type piezoelectric elements. At the frequency of 50 Hz and the acceleration of 0.007 g (for open circuit), the element (PZT 40%) was able to generate up to 50 μV potential (Fig. 8.8a), and up to 40 μV potential for the element with PZT 60% (Fig. 8.8b).

The element (PZT 80%) generated up to 80 μV electric potential. These results were pre-processed by a 500 Hz low-pass filter.

Cantilever-type piezoelectric element with 80% concentration of PZT shows significant results in power generation as a thin layer at low frequencies. Other elements had no signs of ability to convert electrical potential into mechanical energy. The element with PZT 80% was investigated using interferometer PRISM of electronic speckle pattern (Fig. 8.9).

The designed element (PZT 80%) was excited by a sinusoidal function with an amplitude of 5 mV at a frequency of 13 Hz (Fig. 8.8). At the first resonant frequency the

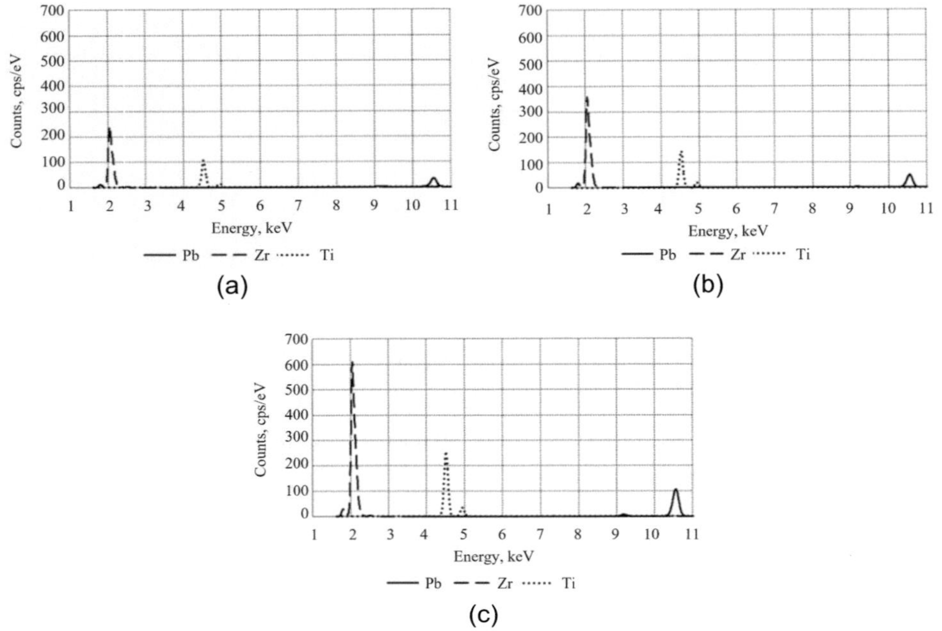

Fig. 8.7 ED spectrum showing peaks of Pb, Zr and Ti of piezoelectric elements with PZT: **a** 40%; **b** 60% and **c** 80%

Table 8.1 Peak values of Pb, Zr and Ti of designed piezoelectric elements

Concentration (%)	Pb 10.5515 keV	Zr 2.04236 keV	Ti 4.50486 keV
40	34.49	244.44	107.73
60	48.62	359.21	143.94
80	108.05	609.85	256.81

element vibrates as a clamped-free cantilever resonator in its fundamental flexural mode. Significant advantage of this element is an ability to apply designed novel piezoelectric composite material at any thickness, form and size on any uniform or non-uniform vibrating surface.

Periodical microstructure and SPR. A novel cantilever-type piezoelectric element (PZT 80%) works in both, direct and indirect, piezoelectric regimes. The designed novel piezoelectric composite is a promising material for future experimentations. This unique property allows a real time and direct observations of affinity interactions, i.e. sensing elements with piezoelectric effect employs the active method for measurements in medical or pharmaceutical fields.

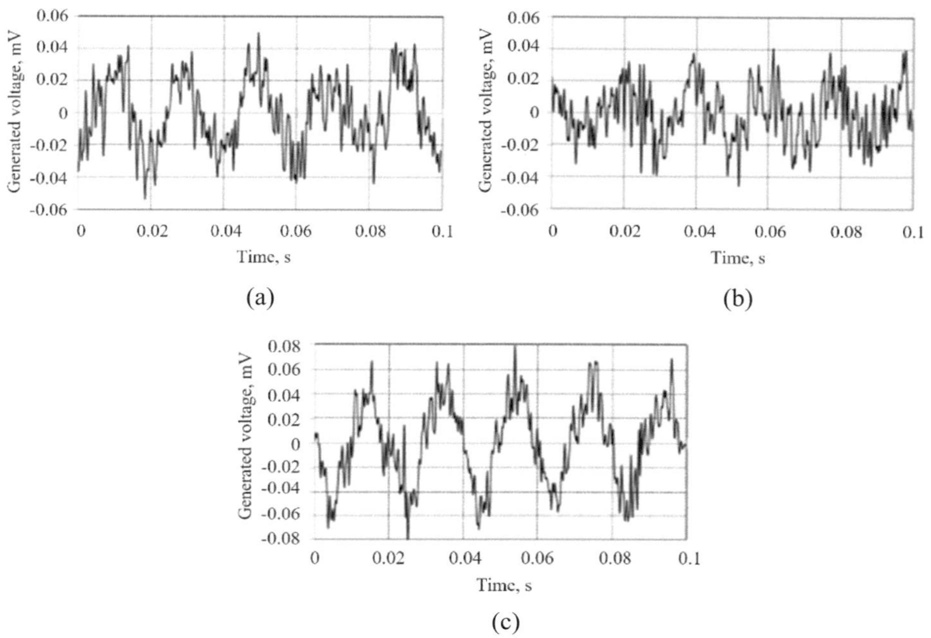

Fig. 8.8 Results of electric potential generated by designed elements with PZT: **a** 40%; **b** 60% and **c** 80%

Fig. 8.9 Hologram of the cantilever-type piezoelectric element with PZT 80%

Imprint of periodical microstructure enables to use this platform for studying biomolecular interactions, to analyze functional information, i.e. the information related to physiological effect of an analyte on a living system [8–10]. It is essential in many important applications: medicine, pharmacy, cell biology, environmental measurements, etc. For this purpose, a four microns periodical microstructure was imprinted on formed thin

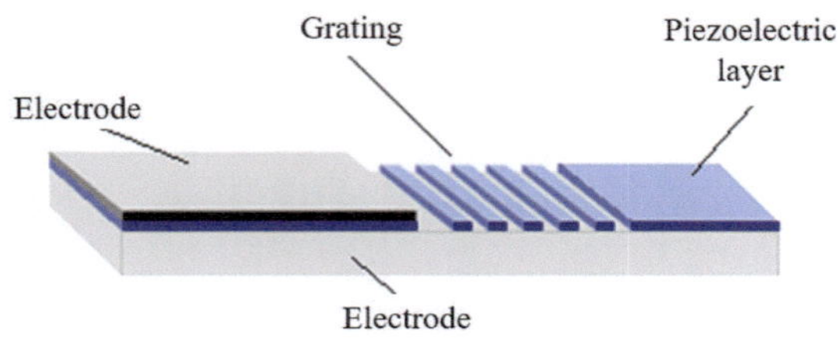

Fig. 8.10 A principal scheme of a cantilever-type piezoelectric sensing platform

film, designed from 80% concentration of PZT (element, exhibiting best piezoelectric properties). Schematic view of the designed element is given in Fig. 8.10.

The platform consists of a thin piezoelectric film coated on a copper foil working as an electrode. Opposite electrode is formed on the top of a thin film. Periodical microstructure was formed by hot-embossing procedure at defined conditions.

Surface morphology of imprinted grating was measured using AFM. As previous researches showed, it is rather hard to imprint periodical microstructure in piezoelectric layer because of its brittleness and inelasticity. Here, an additive polyvinyl butyral was chosen to improve these properties. Thus, a well-defined grating was formed (Fig. 8.11).

Average grating depth was ~580 nm with an average period of 3.8 μm and rather smooth surface–surface roughness R_q = 129.8 nm. Technological data for formation of periodical microstructures are presented in Table 8.2.

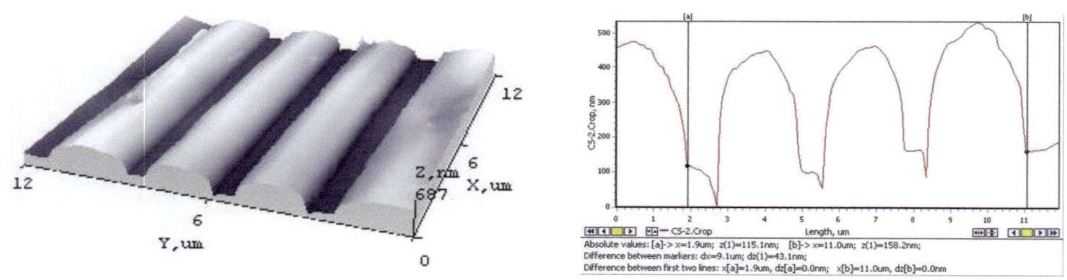

(a) (b)

Fig. 8.11 AFM view of a four-micron periodical microstructure imprinted on a novel piezoelectric element **a** a 3D grating view; **b** topography cross section of the grating

Table 8.2 Technological data for the formation of periodic microstructures

Master grating periodicity	Depth	580 nm
	Periodicity	3.8 μm
Master grating dimensions	Length	3 mm
	Width	18 mm
Grating lines	Parallel to the short edge	
Pressure in metallic mandrel	12 MPa	

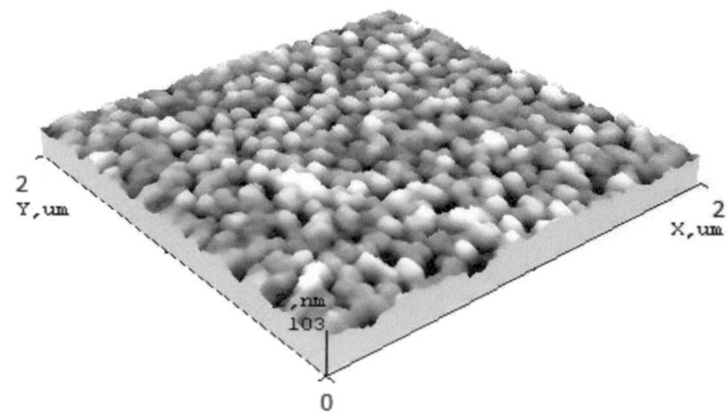

Fig. 8.12 AFM view of a grating top coated with Ag nanoparticles

Main drawbacks of such a sensing platform arise because of the interaction of bio-materials in the investigated analyte and are strongly influenced by the adsorption of bio-molecules, their diameter size and viscosity of the analyte. To overcome these draw-backs, silver nanoparticles were precipitated on the grating. Previous research showed that incorporation of Ag highly increases the optical response of SPR. The SPR enhances the absorption (optical signal); and nanoparticles are used as biological tags for quantitative detection of bio-molecules in analyte. Moreover, a combination of piezoelectric and SPR properties in a single element is an effective way to expand the working range of the element. To prove the relevance of Ag nanoparticles in the designed cantilever-type sensing platform diffraction efficiency measurements were taken (Fig. 8.12).

AFM surface view of the grating (Fig. 8.12) showed that approximate size of Ag was ~ 17 nm diameter.

Diffraction efficiency measurements were performed using a laser diffractometer (Fig. 8.13). A He–Ne red laser light was incident to the grating imprinted on a designed element (PZT 80%). Most of the diffraction energy was concentrated on its zero order (~ 54%) and in its first orders (~ 33%). For a periodical microstructure with silver nanoparticles, diffracted energy in its zero order has dramatically decreased to ~28% and in its first orders of its maximum increased up to 35%. In second orders of maximum, it increased up to 29%. Thus, results imply that silver nanoparticles significantly improve the optical response of a novel sensing platform.

A future perspective of the designed novel cantilever-type piezoelectric element is related to its integration in MEMS for analysis of functional information as physiological effects of an analyte, type and concentration of molecules, and etc. When, for example, a constant potential is applied, electrochemical reactions may be observed together with the changes of mass or resonant frequency shifts. These measurements are often desired in biomedicine, cell biology, and environmental or pharmacy measurements.

Subchapter conclusions. The designed cantilever-type sensing element made of novel piezoelectric material is able to possess resonance frequencies and leads to higher sensitivity and faster response time. The element works at low frequencies and is able to generate up to 80 μV at 50 Hz frequency. It also works vice versa, i.e., at frequency of

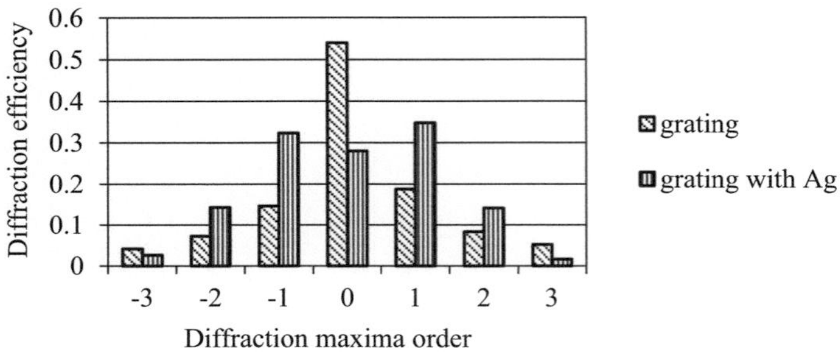

Fig. 8.13 Diffraction efficiency measurement results

13 Hz with the amplitude of 5 mV and induced by a sinusoidal function it vibrates at the first resonant frequency. The designed element vibrates as a clamped-free cantilever resonator in its fundamental flexural mode.

To obtain a sensing platform suitable for medical applications in MEMS, a 4-µm periodical microstructure was imprinted. To improve the sensitivity and diffraction efficiency of the designed element silver nanoparticles of ~17 nm diameter were precipitated on the surface of grating. This was a result of appearance of SPR effect. Incorporation of Ag nanoparticles increase the diffraction efficiency of the grating by 15–18% in its first orders, and about 10% in its second orders of maxima. And property of piezoelectricity enables to tune the grating, i.e., to change the grating parameters (grating width and depth) and its optical properties when voltage is applied.

Results show that designed novel cantilever-type sensing platform exhibits piezoelectric and plasmonic properties. These features are extremely important when designing valuable analytic MEMS instruments for applications such as biomedicine, pharmacy, environment and chemistry. Future experimentations will be related to a practical use of the designed elements for bio-sensing investigations.

8.2 Microfluidic Channels for Bio Particles Manipulation

The concept of an active microchannel for precise manipulation of particles in biomedicine is reported in this subchapter. The novel vibration-assisted thermal imprint method for effective formation of the microchannel network in the nano composite piezo polymer layer is proposed. Different wave length bulk acoustic waves excited in imprinted microstructure enable its functioning in trapping, pattering, valve, or free particle passing modes. Acoustic waves were excited by using a special pattern of electrodes formed on its top surface and a single electric ground electrode formed on the bottom surface. The microchannel development process was started from the synthesis of lead zirconate titanate nanopowder [Pb $(Zr_{0.52}Ti_{0.48})$ O_3]. Further mixing it with three different binding materials (polyvinyl butyral (PVB), poly (methyl methacrylate) (PMMA) and polystyrene (PS)) in benzyl alcohol screen-printing paste was prepared. Then using conventional screen-printing techniques, three types of PZT coatings on copper foil substrates were obtained. To improve the voltage characteristics, the coatings were polarized; their structural and chemical composition was analyzed by scanning electron microscope. Mechanical and electrical characteristics were determined using the *COMSOL Multiphysics* software (*COMSOL*®, Burlington, United States) model with experimentally obtained parameters of periodic response of the layered copper foil structure in it. The hydrophobicity properties of the PZT composite were analyzed by measuring the contact angle between the distilled water drop and the polymer of different composition: PZT with polyvinylbutyral (PVB), poly (methyl methacrylate) (PMMA), and polystyrene (PS). Finally, behavior of the microchannel formed in nanocomposite piezo polymer was simulated applying electrical excitation signal on the pattern of electrodes and analyzed experimentally by the method of holographic interferometry. Wave-shaped vibration forms of the microchannel were obtained that enabled particle manipulation [11]. Microfluidic systems with networks of microchannels that had dimensions from tenths to hundredths of micrometers were used to manipulate the laminar picoliter level of fluid flow, discrete fluid droplets, particles, or cells [12–14]. Their functional, analytical, and sampling features were attractive for applications in (biology), life sciences like cytometry, diagnostics by lab on chip devices, microscale cell culture, etc. [15–18]. At first, microchannels of such systems were made in solid substrates (silicon, glass) [19–23], then, with the development of new plastic materials and their fabrication technologies, polymethylmethacrylate (PMMA), polystyrene (PS), polycarbonate (PC), cyclic olefin copolymer (COC) was started to be used due to their biocompatibility and more effective fabrication [22, 24–29].

The concept of an active microchannel, i.e. the microchannel formed in piezo polymer, was proposed. Thus, advantages of plastic materials and functional ability of the channel to transport droplets due to wave type deformation of its walls were combined.

Acoustophoresis, the method based on application of acoustic radiation force for precise manipulation of pico scale fluid droplets or particles is well known in the world. Especially promising is the application of surface acoustic waves (SAW). Due to the effect of standing acoustic waves (SSAW), particles are patterned across the microchannel, while due to the effect of traveling surface acoustic wave streaming of the particles or droplets is obtained [30–34].

Typically, SAW [30–33] is excited on a $LiNbO_3$ substrate by applying an AC signal on a pattern of electrodes formed on the substrate surface. In [24] integration of a thin PZT film with a silicon/glass substrate was reported into the microchannel structure and generation of bulk acoustic waves (BAWs) for patterning of particles across the microchannel.

In the reviewed cases, the microchannel formations were composed of separate piezo elements ($LiNbO_3$ substrate [30–43] or PZT [34, 35]) and non-active glass, silicon, or PDMS elements. Sources from the reviewed literature do not provide data on the surface microforms or spatial formations of piezo polymer material used in microfluidics. Such formations under the effect of electric excitation at certain zones would enable bidirectional (along and across microchannel) patterning and streaming of the particles.

In the publication [36] the analysis of the development of today and future nano/microfluid mechanical systems clearly shows that the research problem is relevant and needs serious research attention.

In this work, the results of the development of an active microchannel formed in piezo polymer nano composite material are presented by the vibration-assisted thermal imprint process. The schematic view of the developed microfluidic chip with a network of microchannels is shown in Fig. 8.14.

In close-up, a fragment of the channels, imprinted in PZT nanocomposite was highlighted. Analysis of its performance in particle trapping, valve and free pass modes was carried out.

The sub-chapter is organized as follows: first, synthesis of the PZT composite material and fabrication of the PZT composite coating to be used for microchannel network formation is presented. Furthermore, an analysis of its composition and a determination of its mechanical characteristics are performed. Then hydrophobicity analysis of PZT coatings interacting with a water droplet is carried out; next vibration assisted thermal imprint method is discussed and characteristics of the microchannels obtained by it are presented.

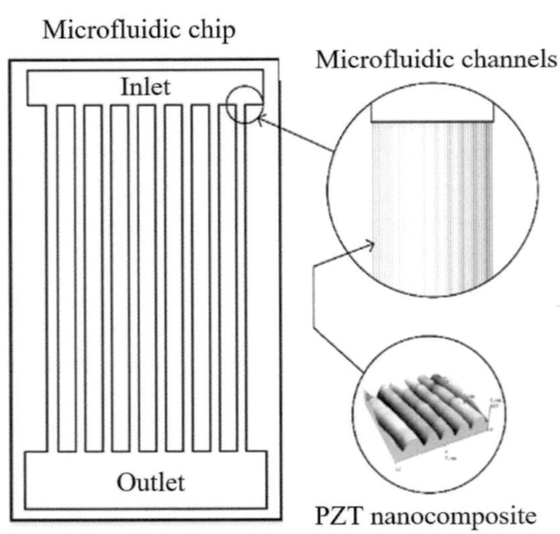

Fig. 8.14 Schematic view of a microfluidic microchip

Finally, the simulation of a single microchannel behavior under periodic excitation is carried out by *COMSOL Multiphysics* software (*COMSOL*®, Burlington, United States) and its performance is experimentally tested.

Materials and fabrication process. *Synthesis of PZT composite material.* Lead zirconate titanate nanopowder [Pb (Zr_x, Ti_{1-x}) O_3] was synthesized applying the method of an oxalate/hydroxide co-precipitation using the following materials: lead (II) acetate $Pb(NO_3)_2$, titanium butoxide $Ti(C_4H_9O)_4$, zirconium butoxide $Zr(OC_4H_9)_4$ (80% solution in n-butanol), oxalic acid dihydrate $C_2H_2O_4 \cdot 2H_2O$, 25% ammonia solution and deionized water. In deionized water (100 ml), 26 g of lead (II) acetate was dissolved. In a separate glass with 500 ml of deionized water oxalic acid dehydrate (32 g) was dissolved followed by heating the solution up to 50 °C, afterwards adding titanium butoxide 5.1 g and 80% zirconium butoxide solution 7.65 g by drops. Then, the mixture was stirred intensively to obtain a clear yellow solution. This titanium and zirconia alkoxides solution was mixed with lead acetate solution and alkalizing with 25% ammonia solution till pH 9–10 was achieved continuing to mix the solution for 1 h. The precipitate of the white amorphous PZT precursors was filtered in vacuum and washed with deionized water and acetone during filtering. Then the material was dried at 100 °C for 12 h. The obtained powder was calcinated for 9 h at 1000 °C. In the final stage, the PZT powder was milled and mixing it with 30% solution of binding material (polyvinyl butyral (PVB), poly (methyl methacrylate) (PMMA) and polystyrene (PS)) in benzyl alcohol screen-printing paste was prepared. The ratio of the components was defined to be 80% of PZT and 20% of the binder in a dry coating (0.83 g of 30% PVB, PMMA or PS solution for 1 g of PZT powder). The paste viscosity was adjusted to 40 ± 5 Pa s with benzyl alcohol (Brookfield Viscometer, ABZ spindle, 10 rpm, 25 ± 1 °C).

Formation of PZT composite material layer. In the next stage of active microchannel development, piezo polymer coating in which the active channel have been imprinted was formed on a copper foil. Three coatings with different binding materials were formed and investigated: element 1 with PZT/PMMA coating, element 2 with PZT/PS coating and element 3 with PZT/PVB coating.

Using conventional screen printing techniques, the paste was applied to copper foil substrates using a 48/70 mesh polyester monofilament screen. The properties of polyester monofilament screens and the thickness of the PZT composite layer are presented in Table 8.3.

The photos of the polyester monofilament screen and layer applicator are shown in Fig. 8.15.

The coatings were then dried at a temperature of 100 °C for 30 min in an electrical oven. Images of the coating samples on copper foil are shown in Fig. 8.16.

The dimensions of a sample were 55×65 mm.

Polarization of the coating. Further, these samples were polarized to align the dipole vectors of PZT grains in it and obtain the resultant polarization vector perpendicular to the

Table 8.3 Properties of polyester monofilament screens and PZT composite layer thickness

Meshed screen type	32/70	48/70	140/34
Mesh opening, μm	245	130	30
Thread, μm	70	70	34
Open area, %	60.5	42.3	22.0
Mesh thickness, μm	108	107	52
Theoretical ink volume, cm^3/m^2	65	46	11
Formed PZT layer thickness, μm	68 ± 1	60 ± 1	25 ± 1

Fig. 8.15 Photo of: **a** the polyester monofilament screen; **b** layer applicator

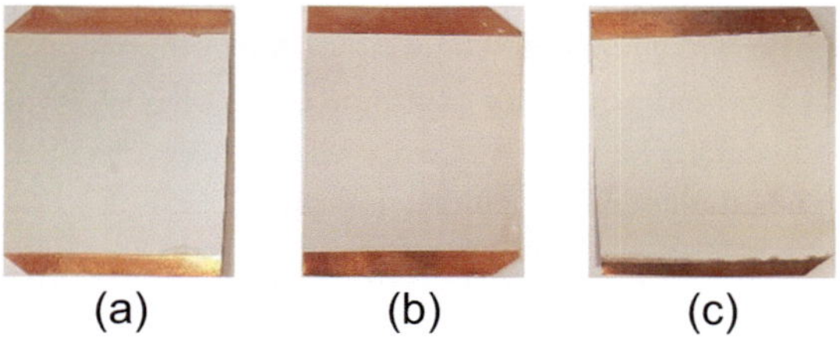

Fig. 8.16 Samples of the coating on copper foil: **a** PZT/PMMA; **b** PZT/PS; **c** PZT/PVB

coating plane. Using the method of vibration-assisted thermal imprint process, the network of active microchannels was formed. Thus, microchannels, being surface special formations in piezo polymer, under the excitation of an electric field undergo deformations due to inverse piezo effect. Polarization of the coating was performed by poling, i.e. electrical pole alignment by high voltage applied on the coating. The copper foil substrate with the coating on it, for 30 min was placed in 5 kV potential by fixing it with the help of special brackets in between positive and negative poles. Under the effect of electrical field, PZT grain poles were aligned, which resulted in improved voltage characteristics of the piezoelectric coating. Pole alignment set is presented in Fig. 8.17.

For the analysis of electrical and mechanical properties, 15×25 mm size samples were cut from a copper foil substrate with poled PZT polymer coating and contact wires were attached to them.

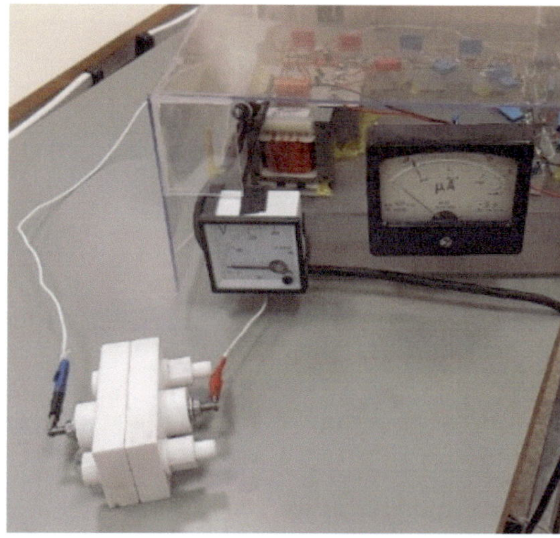

Fig. 8.17 Pole alignment set

Experimental analysis of PZT coatings. *Analysis of structural and chemical composition.* The Quanta 200 FEG scanning electron microscope (SEM) integrated with the energy dispersive X-ray spectrometer (EDS) detector X-Flash 4030 from Bruker (Pubcompare, Switzerland) was used to analyze the structure and chemical composition of the synthesized PZT composite material. In a controlled pressure water steam atmosphere, three poled elements 1, 2, 3 with different binding material were examined. A 133 eV (at Mn K) energy resolution at 100.000 cps was achieved with a 30 mm^2 area solid state drift detector, cooled with Peltier element. The energy differences between the quantum states of the system were measured and the probabilities of system jumps between these states were determined. The method of X-Ray spectroscopy was applicable for the analysis of energy distributions. SEM images of the analyzed samples are presented in Fig. 8.18.

PZT/PMMA PZT/PS PZT/PVB

Fig. 8.18 SEM views of samples with different binders mixed with PZT

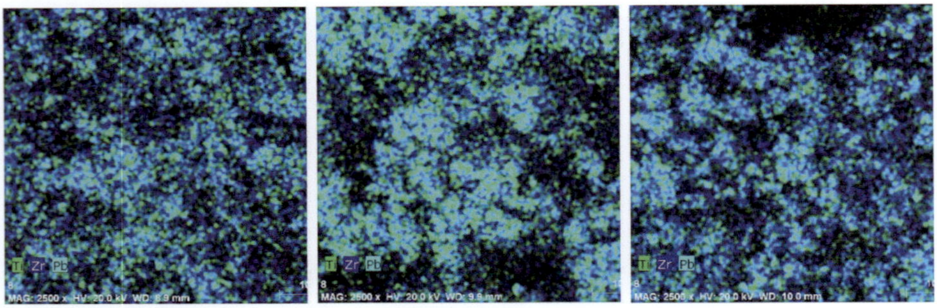

PZT/PMMA PZT/PS PZT/PVB

Fig. 8.19 Elemental mapping done with SEM of the samples with different binders mixed with PZT

The images reveal the granular structure of sample 1 with a grain size of ~1.1 μm in diameter on the surface. The surface of sample 2 was smoother and its grains were smaller of 0.9 μm in diameter. There were 3D structures in sample 3 with empty cavities of 6–8 μm in diameter. The full composition of the elements can be seen in Fig. 8.19.

The main composition elements were carbon (C) and zirconium (Zr); these are both good conductors, which is the condition for good piezoelectric properties of the novel PZT coatings developed.

Mechanical and electrical characteristics of PZT coatings. Mechanical and electrical characteristics of multilayer foil structure with PZT coating were determined using *COMSOL Multiphysics* software (*COMSOL*®, Burlington, United States) model and applying in it the registered experimentally mechanical and electrical parameters of periodic response of the structure. The vibration amplitude was measured with the help of the LK-G3000 triangular laser displacement sensor (Keyence, Mechelen, Belgium) and the electrical potential generated was collected with a PicoScope 3424 USB oscilloscope (Pico, Cambridgeshire, United Kingdom). Experimental setup is presented in Fig. 8.20.

When the first vibration mode was analyzed using a simulation model in the COMSOL Multiphysics software (COMSOL®, Burlington, United States) with the registered data, the mechanical characteristics of each sample were calculated (Table 8.4).

The highest module of elasticity (6.3 GPa) was observed for the PZT nanocomposite with the PMMA binding material, while the usage of PS and PVB leads to lower modulus of elasticity of 5.3 and 3.9 GPa, respectively (Table 8.4). 80% of the nanocomposite material was PZT nanoparticles (Young modulus 63 GPa) while the share of the binding material was 20%. However, because of it, the modulus of elasticity was reduced approximately 10 times. These results were in correlation with the data found in the literature: Young's modulus of PMMA was 3.1 GPa [37], PS—2.7 GPa [38] and PVB had the lowest modulus of elasticity of 50 MPa [39].

Fig. 8.20 Experimental setup for determination of mechanical and electrical characteristics of the samples: 1—sample holder; 2—LK-G82 sensor head; 3—*LK-G3001PV* control block; 4—*PicoScope* oscilloscope; 5—computer

Table 8.4 Mechanical properties

Binding material	ρ, kg/m^3	ζ	f, Hz	ω_n, rad/s	E, GPa
PVB	6298	0.0052 ± 0.0003	26.5 ± 0.4	166 ± 2.8	3.9 ± 0.3
PMMA	6316	0.0077 ± 0.0006	29.4 ± 0.9	185 ± 5.8	6.3 ± 0.8
PS	6288	0.0080 ± 0.0004	28.3 ± 0.6	178 ± 4.0	5.3 ± 0.4

Periodic vibrations of the analysed specimen were excited and electrical potential due to the piezoelectric effect of the PZT material was generated. The voltage generated by periodically excited specimens is presented in Fig. 8.21.

The highest value of the generated electrical potential was registered for the specimen with the PS binding material. The generated voltage reached 3 mV. The lowest result of

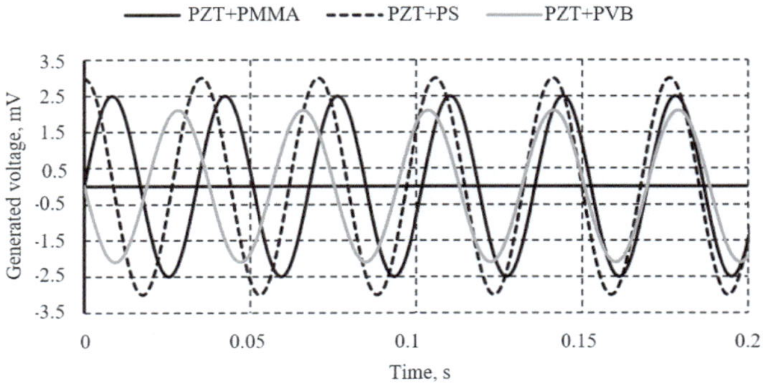

Fig. 8.21 Generated voltage by periodically excited specimens

2.1 mV voltage was recorded for PZT with PVB binding material. The PMMA binding material had a 2.5 mV value (Fig. 8.21).

Hydrophobicity analysis of PZT coatings. The network of microchannels was a spatial formation imprinted on the developed PZT composite. Because it was intended to be used for precise manipulation of droplets; hydrophobic properties of the surface contacting the manipulated fluid were important. The hydrophobic properties of the PZT composite were analyzed by measuring the contact angle at the PZT surface of the interface—distilled water droplet. The structure of the experimental setup for contact angle measurement is shown in Fig. 8.22.

The parts of the experimental setup were placed on the surface of a table which was isolated from external excitation sources. The setup consisted of two optical lenses (focal length 600 mm) placed between the camera (*Guppy F-503 B&W CMOS*, *Allied Vision, Stadtroda, Germany*) and the sample with a water droplet on its surface. The position height of the cameras could be adjusted with respect to the droplet (Fig. 8.23).

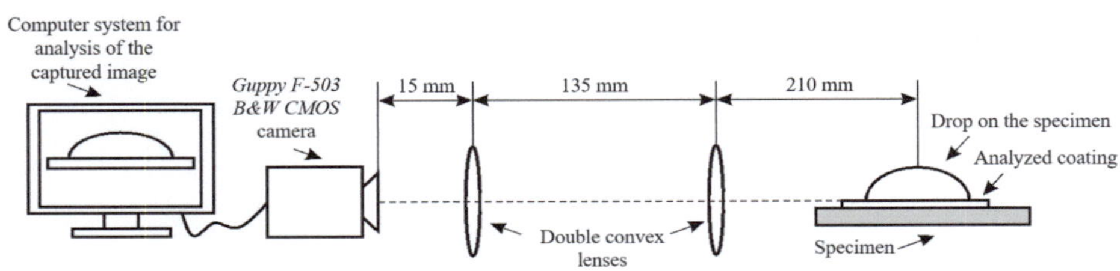

Fig. 8.22 Experimental setup for contact angle measurement of hydrophobic and hydrophilic material

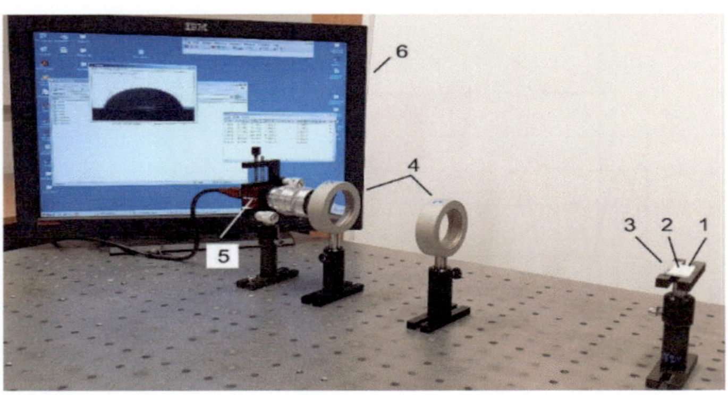

Fig. 8.23 Experimental setup for contact angle measurement: (1) drop on the specimen; (2) analyzed coating; (3) specimen; (4) double convex lenses; (5) *Guppy F-503 B&W CMOS* camera; (6) computer system for analysis of the captured image

Testing was performed in a dark laboratory room with the lights set in which the liquid droplet appeared to be black. This was necessary to ensure measurement accuracy and perform image analysis. For contact angle measurement, samples of PZT composites with three different binding materials, polyvinyl butyral (PVB), poly(methyl methacrylate) (PMMA), and polystyrene (PS), were used. As the liquid, distilled water droplets were used.

First, for accurate positioning of the droplet image, the height of the specimen holder was adjusted, positioning the specimen parallel to the camera vision field. Then a 0.02 μl in volume droplet of distilled water in volume was diffused from the pipet onto the analyzed surface from a height of 1 cm. For image processing, the open-source software ImageJ (National Institutes of Health, JAV) provided by Wayne Rasband was used. The drop snake method implemented in its plugin was applied for contact angle measurement. This method is based on the use of a polynomial fit to obtain the droplet profile curve. 7 knots starting from the left lower end to the right lower end are placed along the profile of the droplet. The droplet profile is then approximated as a polynomial curve through these knots and the contact angle between it and the sample surface is determined.

A water droplet was tested on three different samples of PZT polymer composite, polyvinyl butyral (PVB), polymethyl methacrylate (PMMA), and polystyrene (PS). Each test was performed under the same parameters and the images obtained were analyzed several times. The results are presented in Fig. 8.24.

The graphic representation of the measured contact angle (θ) for different multilayer polymer specimens is shown in Fig. 8.25.

The maximum value of the contact angle was observed with PZT + PMMA. It is 92.94° with an error of ±1°. The lowest contact angle was observed for PZT + PVB—80.71° with a measurement error ±1.3°. The measured value of the contact angle for PZT + PS was 88.8° with a measurement error of ±1.16° and falls between the measured values of PMMA and PVB. According to [40, 41], the polymer material is considered hydrophilic if the contact angle of the water is less than 90°. Therefore, it could be stated

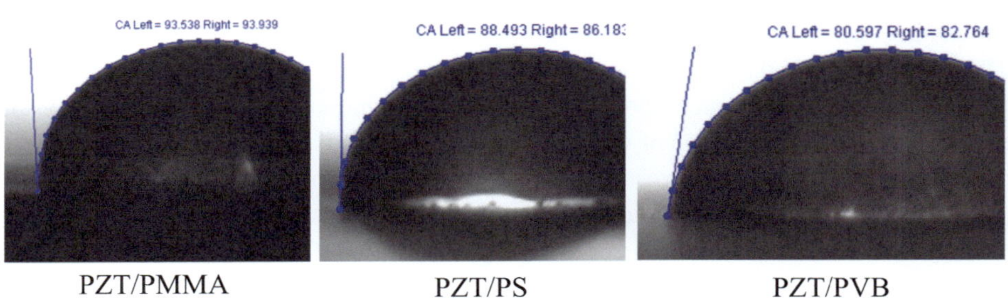

Fig. 8.24 Polynomial fit of the droplet profile through knots starting from the left lower end to the right lower end

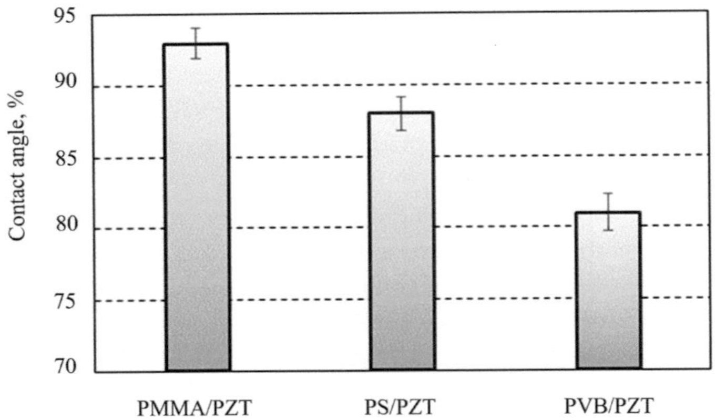

Fig. 8.25 Mean contact angle of a water droplet on the PZT composite surface

that PMMA was in the range of hydrophobic materials with water, while PVB and PS could be considered hydrophilic surfaces with water as their contact angle was less than 90°.

Vibration-assisted thermal imprint of the microchannel network. The parameters of the microchannel network formed by the vibration-assisted thermal imprint method were as follows: length 20 μm, period 4 μm, depth 0.56 μm width of the land 2 μm and width of the ridge 2 μm.

Currently, different microchannel fabrication techniques are known [36]: micro deformation, micro machining, lithography, MEMS (DIRE), laser micromachining. Each of them has its own advantages and shortcomings, e.g. micro-deformation is low-cost, quick, relatively low precision technique after which additional processing is necessary; micromachining is low-cost, sufficient precision technique, nevertheless only simplistic shapes can be machined; lithography is slow process technique, laser micro-machining is relatively expensive technique though it can give good accuracy, shape complexity, process rate parameters.

The proposed method of vibration-assisted thermal imprint combines advantages of several techniques mentioned above—it is the fast technique like micro-deformation but enables much higher precision of the obtained replica; and gives the possibility to replicate more complex geometrical forms. A general view of the kit for vibration-assisted thermal imprint of microstructures (or microstructure replication) is shown in Fig. 8.26.

It includes the Tinius Olsen material testing machine (H10KT, Tailored Test Solutions Ltd., Pennsylvania, USA) (1), with fixed replication unit, Agilent 33220A function waveform generator (Keysight, Santa Rosa, JAV), linear amplifier EPA-104 (Piezo Systems Inc., Woburn, MA, USA), heating circuit power source (4), PC with the software installed for signal processing (5). The main functional part, the replication unit that is fixed in the

Fig. 8.26 Setup for vibration assisted thermal imprint of microstructures: (1) testing machine *Tinius Olsen*; (2) function waveform generator *Agilent 33220A*; (3) linear amplifier *EPA-104*; (4) power source; (5) PC with the installed software for signal processing; (6) locking ring; (7) piezo stack; (8) heating element; (9) cubic shape front mass

grip elements of the testing machine. Here, the sonotrode above the PZT polymer-coated copper foil can be clearly distinguished. Detailed structure of the sonotrode is given, too. On the bottom surface of the cubic shape front mass (9), a nickel master matrix is attached. High frequency excitation is applied to the front mass with the help of a piezo stack (7) attached to its top surface, and it is heated using heating element (8) perpendicularly traversing it. The dimensions of the cube are $22 \times 22 \times 22$ mm; the piezo stack consists of six piezoceramic rings (piezoceramics—PZT-5H) with the following dimensions: outer diameter 22 mm, inner diameter 10 mm, thickness 5 mm. The imprint process is carried out as follows—the sonotrode with the attached mass matrix is pressed against the surface of copper foil with the PZT polymer coating with the force creating 0.5 bar pressure at the contact under 11 kHz excitation generated with the help of piezo stack and 1480 °C temperature achieved using the heating element. After maintaining the pressure for 10 s, replica of the microstructure is obtained. The geometrical characteristics of its profile are shown in Fig. 8.27.

Numerical modeling and experimental analysis of microchannel performance. The behavior of the imprinted microchannel was investigated numerically using COMSOL

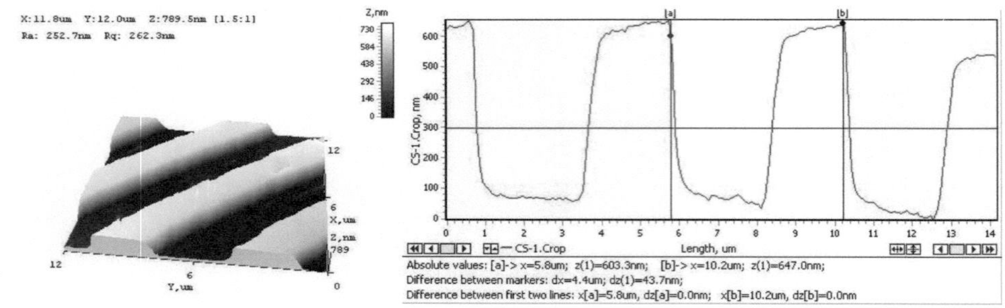

Fig. 8.27 Atomic force microscope image and cross section of the replica: profile—lamellar, period—4 μm, land—2 μm, ridge—2 μm, depth—560 μm

Multiphysics 5.2 software (COMSOL®, Burlington, United States). The microfluidic system was periodic. It consisted of several parallel microchanels of the same geometry ($L = 20$ μm, $P = 4$ μm, $T = 2$ μm, $d = 0.56$ μm, $w = 2$ μm, $2r = 2$ μm) connected to the outlet and inlet fluid containers. Therefore, in the model, only one microchannel was analyzed. Symmetrical boundary conditions were applied: the left and right surfaces were fixed using rollers, and the front and back surfaces were fixed unmovably as they were attached to fluid containers. The bottom surface was attached to other components of the sensor; therefore, it was fixed fully and served as an electrical ground. Electrodes were formed on the top of the microchannel system. It was covered by a transparent layer that allowed visual inspection of the fluid flow and ensured high pressure inside the channel. Vibrations in the network of microchannels were generated by a sinusoidal electrical signal of 20 V. The dimensions and boundary conditions of the finite element model of the microchannel are presented in Fig. 8.28.

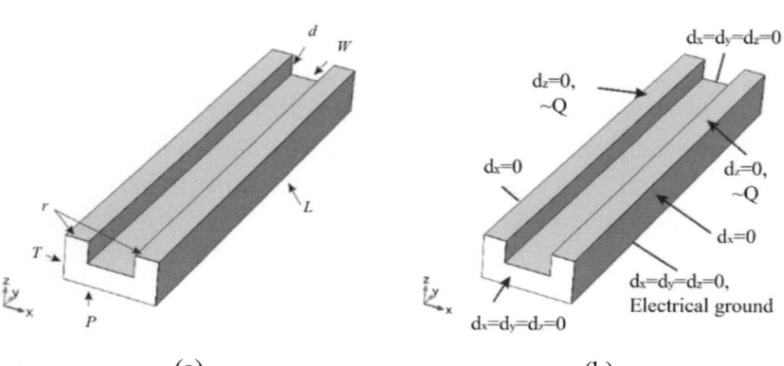

Fig. 8.28 The finite element model of microchannel: **a** dimensions; **b** boundary conditions

The results of the dynamical response simulation of the microchannel are presented in Figs. 8.29, 8.30 and 8.31.

The deformation of the system electrically generated at the 110 MHz frequency—the first natural frequency is shown by the displacement field of components X (Fig. 8.29). It can be observed that the microchannel oscillates in half-wave mode—the cross-section area increases or reduces in phase through all the length of micro channel; therefore the

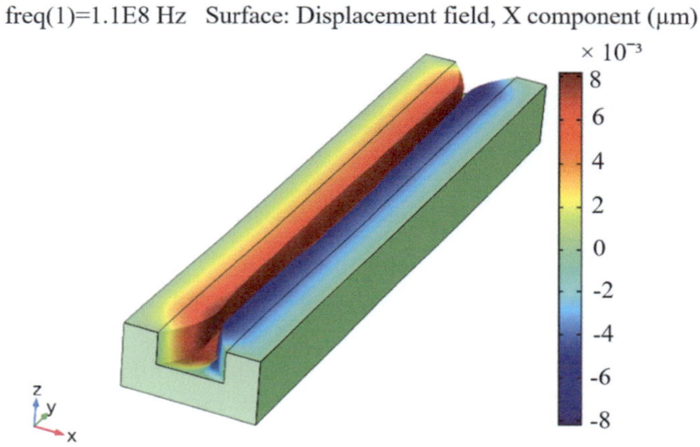

Fig. 8.29 Surface displacement field of the microchannel periodically excited at a frequency of 110 MHz—X component visualization

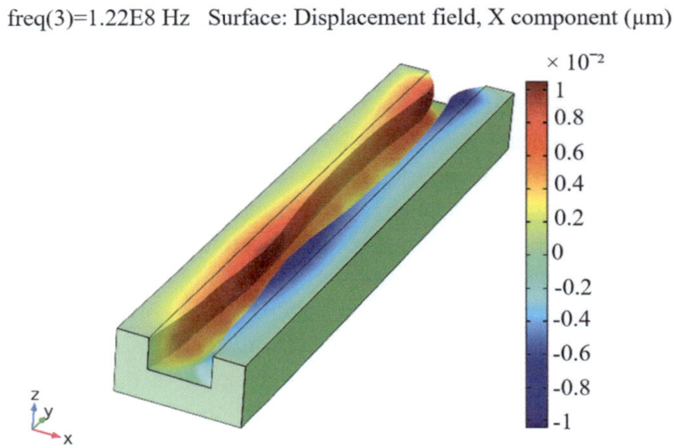

Fig. 8.30 Surface displacement field of the microchannel periodically excited at the frequency of 122 MHz—X component visualization

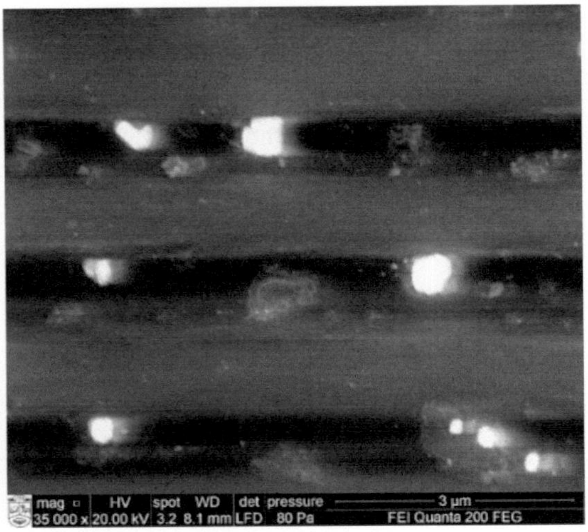

Fig. 8.31 SEM image of bioparticles in motion

micro channel functions as mechanical valve, bio particles cannot pass it, i.e. they are stopped near the entrance of it. The system excited at the frequency of 122 MHz, the second natural frequency of the system oscillates in full wave mode, and the two halves of the micro channel along its length oscillate with 180 shifted in phase (Fig. 8.30). Thus, the channel is divided into two segments with different concentrations of bioparticles: particles trapped at midpoint (wave node) and particles free at both ends (Fig. 8.30). In the case of no oscillations of the channel, the particles can freely pass it.

The channel function in different modes of particle manipulation was analyzed using SEM Quanta 200 FEG (Pubcompare, Switzerland). SEM images of bioparticles transport are presented in Fig. 8.31. The first row represents the trapping of the particle at the midpoint of the channel; the second row free pass of the particles and the third row represents the case of mechanical valve.

The case of pressure-driven microflow was used for determination of flow characteristics in the microchannel network. The microchannel network analyzed was a periodic microfluidic structure (period 4 μm, depth 0.56 μm) which consisted of 100 microchannels. The pressure of 1 bar was maintained in the inlet container. Microflow analysis was performed with three types of liquid: pure water, acetone, and glycerol. In all cases, the calculated Reynolds number was significantly below 1. If the Reynolds number is very small, much less than 1, then the fluid will exhibit Stokes, or creeping flow, where the viscous forces of the fluid dominate the inertial forces. The parameters determined for pressure-driven microflow for different functioning modes of the microchannel are presented in Table 8.5. The experiment with the application of glycerol was not successful, as the flow rate obtained was very low and it was impossible to measure it. The flow rate values obtained were in correlation with the vibration shapes of the microchannel. When the channel vibrated at the first natural frequency of 110 MHz, an almost uniform change

Table 8.5 Pressure-driven microflow of different modes

Fluid		Pure water	Acetone	Glycerol
Viscosity, cP		1	0.32	1400
Density, g/cm^3		1	0.79	1.26
Theoretical flow rate, μl/min		0.145	0.452	0.0001
Experimental flow rate, μl/min	free flow	0.11 ± 0.002	0.42 ± 0.004	–
	Valve mode, f = 110 MHz	0.05 ± 0.001	0.2 ± 0.002	–
	Trapping mode, f = 122 MHz	0.08 ± 0.001	0.31 ± 0.003	–

in its cross-section area was observed throughout its length. When the channel vibrated at the second natural frequency of 122 MHz, the most expressed change in cross-section area was in certain zones of the channel. Thus, the highest flow rate was obtained when no vibrations were generated in the channel, as there was no reduction of the cross-sectional area. In case of the first vibration mode, the flow rate was the lowest. This corresponded to the highest change in the cross section area. Vibrations at the second natural frequency were the intermediate case both from the viewpoint of cross section area change and the flow rate. The results of the pressure-driven microflow of different modes are presented in Table 8.5.

The numerical-experimental laser interferometric method (was selected) can be applied for identification of liquid (concentration) flow characteristics in the periodic microstructure. The optical properties of the microfluidic were investigated by a non-destructive optical laser diffractometer. He–Ne laser diffractometer ($\lambda = 633$ nm) was used in order to register intensities of reflected or transmitted diffraction maxima efficiencies. All liquids were characterized by relative diffraction efficiencies. Relative diffraction efficiency RDE was defined as the ratio of intensity of diffracted light to the i-th diffraction maxima ($0, \pm 1, \pm 2$, etc.) with intensity of the reflected light from the surface without micro relief. Liquids' refractive index was determined from comparison of theoretical and experimental diffraction efficiencies. The optical properties of the materials used in the calculations are presented in Table 8.6.

Diffraction efficiencies of the periodic microstructures without an analyte (in air or vacuum) showed that periodic microstructure (period 4 μm, depth 0.56 μm) was designed for operation in reflecting mode, i.e. first order diffraction maxima were three times higher than zero order diffraction maximum (Fig. 8.32). Therefore, the periodic microstructure in the reflecting mode was sufficiently sensitive to the refractive index (density) of the analyte. The refractive index changed from 1.33 (pure water) to 1.47 (glycerol) leading to the decrease in the diffraction efficiency of the first order maxima four times from 24 to 6% (Fig. 8.32).

Table 8.6 Optical properties of the materials used in calculations

Materials	Application	Refractive index	Extinction coefficient
Air	Superstrate	1	0
Aluminum (Al)	Substrate	1.4495	7.5387
Pure water (H_2O)	Testing liquid	1.3317	0
Acetone (C_3H_6O)	Testing liquid	1.3578	0
Glycerol ($C_3H_5(OH)_3$)	Testing liquid	1.4707	0

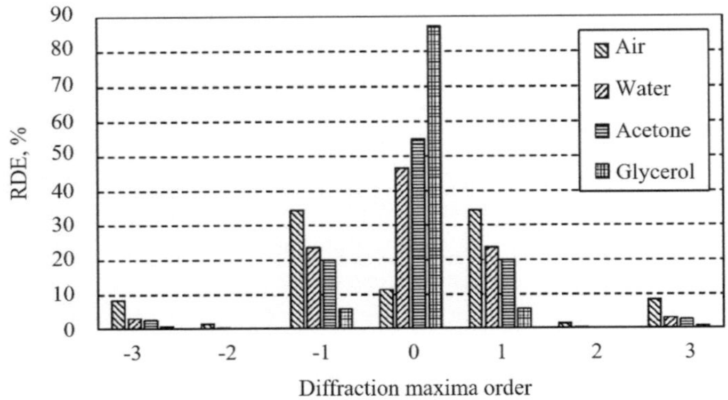

Fig. 8.32 Relative diffraction efficiencies of the reflected light from the periodic microstructure coated by Al

Results confirmed the idea of liquid concentration identification in the periodic microstructures applying numerical-experimental laser interferometric methods.

Subchapter conclusions. In the subchapter, the results of research and development of an active microchannel for improving the effectiveness and accuracy of bioparticle manipulation by acoustic methods were presented.

In conclusion, it can be stated that the main results of the research were as follows:

1. PZT nano composite polymer with elasticity modulus of 6.3 GPa for the PZT nanocomposite with PMMA binding material; 5.3 and 3.9 GPa, respectively, for PS and PVB binding materials: which was suitable for acoustic field generation.
2. The hydrophobicity (hydrophilicity) properties experimentally determined by measuring the water contact angle on its surface revealed that, properly selecting the binding material of the compound, the hydrophobic or hydrophilic polymer surface necessary for bio particle manipulation could be obtained: for PZT with polyvinyl butyral (PVB), poly (methyl methacrylate) (PMMA) or polystyrene (PS) contact angle was in

the range from 92.94°, (for PZT + PMMA) to 80.71°, (for PZT + PVB), which allows stating that PMMA is hydrophobic material with water while PVB and PS showing contact angle less than 90° could be considered hydrophilic surfaces with water.

3. The proposed method of vibration-assisted thermal imprint—thermal replication of microstructures with high frequency assistance enabled formation of a microchannel network of the necessary geometry and precision for bulk acoustic wave generation in the developed PZT nano composite.
4. Performance of microchannel simulated by COMSOL software (COMSOL®, Burlington, United States) showed that it can function in three modes: closed for bio-particle valve mode when half-wave length oscillations of 110 MHz frequency were generated; bio-particle patterning or trap mode when full-wave length 122 MHz frequency oscillations were generated; free pass for bio-particles mode when no vibrations were generated in the microchannel.
5. The theoretical flow rate values in the microchannels were in agreement with the experimentally obtained flow rate values in the case of pressure driven microflow when no vibrations were generated, for example, the theoretical value for water was 0.145 μl/min and the experimentally obtained value was 0.11 μl/min when the pressure of 1 bar was maintained in the outlet container. The change in cross-sectional area of the channel when the first mode 110 MHz frequency vibrations were generated and the second mode 122 MHz frequency vibrations gave a flow rate reduction of up to 0.05 and 0.08 μl/min accordingly (for water).

8.3 Identification of Liquid Concentration in Periodic Microstructures

The investigation of high-frequency vibrations of the fluid is an important problem in the design of microfluidic devices (periodic microstructures) for bioengineering applications. Laser interferometric methods such as time average holography, high speed double expose holography or laser light diffraction allow to do analysis of high-speed fluid flow and dosing or vibration of micro components used in biological and chemical microsystems. Methodology for identification of liquid concentration in the periodic microstructures applying numerical-experimental laser interferometric methods is presented in this subchapter.

Introduction. The main component of a lab-on-a-chip is periodic microstructure. Micro channels (periodic microstructures) are used for the fluid transportation, mixing, separation, and other processing. The set of micro channels is formed by soft-lithography, photopolymerization [42, 43], holographic lithography [44], laser ablation [45], micro-injection molding [46, 47], phase separation [48], gas foaming [49] or 3D printing [50] using a wide range of materials such as polydimethylsiloxane, silicon, polycarbonate, glass, rubber, aluminum and others [51–53]. The use of new materials, such as piezoelectric nanocomposites, allows the development of new periodic microstructures with new features such as piezoelectric properties or specific optical characteristics. Control methods of microfluidics by applying acoustic manipulation are known in the world. For this purpose, standing and traveling waves are excited in microchannels [54]. The generated walls of micro hydrodynamical systems are excited by vibration methods that ensure a more effective flow of micro fluids [55].

The investigation of high-frequency vibrations of the fluid is an important problem in the design of microfluidic devices. Laser interferometric methods such as time average holography, high-speed double-exposure holography or laser light diffraction allow analysis of high-speed fluid flow and dosing or vibration of micro components used in biological and chemical microsystems [56]. Therefore, the methodology for the identification of liquid concentration in the periodic microstructures applying numerical-experimental laser interferometric methods is proposed in this subchapter.

Main equations of holography for phase object investigation. The two-dimensional flow of the ideal compressible liquid in the microchannel (depth is h) is illuminated by laser (wavelength—λ). The phase change of the light of the laser beam that goes through the fluid is [57]:

$$\varphi(x, y, t) = \frac{2\pi}{\lambda}\left[n_{flow}(x, y, t) - n_S\right]h \tag{8.1}$$

where n_S is the refractive index of fluids under static conditions and n_{flow} under flow. The refractive index of the liquid in flow condition depends on the dynamic $\rho_{flow}(x, y, t)$ and static ρ_s densities and could be expressed as:

$$n_{flow}(x, y, t) = 1 + \beta\frac{\rho_{flow}(x, y, t)}{\rho_s} \tag{8.2}$$

where the correction coefficient β should be determined from them experiments.

If time average hologram is recorded of that light during a time interval $T \gg \frac{1}{\omega}$, then phase factor of the reconstructed wave is proportional to:

$$u(x, y) = \frac{1}{T}\int_0^T \exp\left[J_0(\varphi(x, y, t))\right]dt \tag{8.3}$$

The intensity of the observed interference pattern is proportional to:

$$I(x, y) = |u(x, y)|^2 \tag{8.4}$$

Also, it is possible to use high speed double exposure holography for the fluid flow analysis if the flow varies steadily and the exposure time is $T \ll \frac{1}{\omega}$. Then, it could be assumed that the density field is constant during the exposure. For this case, the interference pattern is proportional to:

$$|u(x, y)|^2 = \cos^2[\overline{\varphi}(x, y)/2] \tag{8.5}$$

where $\overline{\varphi}(x, y)$ is the change in phase between the two exposures.

Determination of the concentration of liquids. For the development of methodology for the identification of liquid concentration in the periodic microstructures applying numerical-experimental laser interferometric methods, the periodic microstructure (period 4 μm, depth 0.7 μm) made from aluminum and polycarbonate and three liquids (pure water, acetone and glycerol) was selected. The optical properties of the selected materials illuminated by a red laser ($\lambda = 633$ nm) are presented in Table 8.7.

Optical properties of the microfluidic were investigated by a nondestructive optical laser diffractometer (Fig. 8.33).

The He–Ne laser diffractometer ($\lambda = 633$ nm) was used to register the register the intensities of the reflected or transmitted diffraction maxima efficiencies. All liquids were characterized by relative diffraction efficiencies. The relative diffraction efficiency RDE was defined as the ratio of the intensity of diffracted light to the first diffraction maxima (0, ± 1, ± 2, ...) with the intensity of the reflected light from the surface without micro relief [58]. The liquid density could be calculated from the measured refractive index.

Table 8.7 Optical properties of the materials used in thein the calculations

Materials	Application	Refractive index	Extinction coefficient
Air	Superstrate	1	0
Aluminum (Al)	Substrate	1.4495	7.5387
Polycarbonate (PC) ($C_{16}H_{14}O_3$)	Substrate	1.5805	0
Pure water (H_2O)	Testing liquid	1.3317	0
Acetone (C_3H_6O)	Testing liquid	1.3578	0
Glycerol ($C_3H_5(OH)_3$)	Testing liquid	1.4707	0

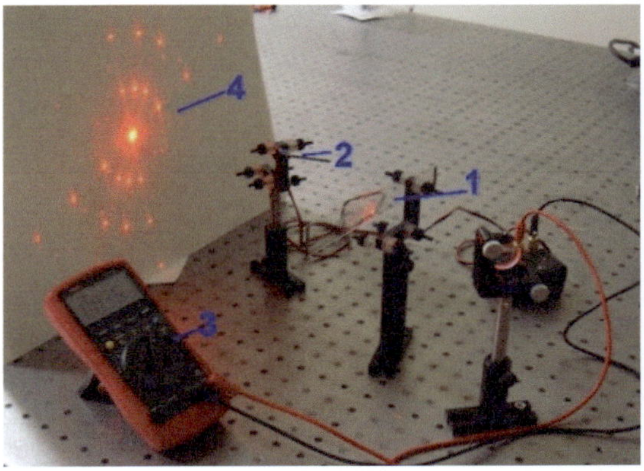

Fig. 8.33 Measuring schema of diffraction maxima: 1—sample; 2—photodiode; 3—ampere meter; 4—maxima distribution

The diffraction efficiencies of the microfluidic illuminated by laser were theoretically calculated using the commercial software GSolver V5.2 (Grating Solver Development Company, Saratoga Springs, JAV). The GSolver V5.2 software for the calculation of microfluidic optical parameters is presented in Fig. 8.34.

GSolver (*Grating Solver Development Company,* Saratoga Springs, JAV) is based on the modal method. It is set up to work with linear isotropic homogeneous materials and is a full vector implementation of a class of algorithms known as Rigorous Coupled Wave (RCW) analysis.

Results. Two periodic microstructures fabricated in aluminum (Al) and polycarbonate (PC) were investigated numerically and experimentally with three liquids (pure water, acetone, and glycerol) to identify the sensitivity of the proposed method. For the measurement of diffraction maxima of the reflected light, Al microstructure was used, while maxima of

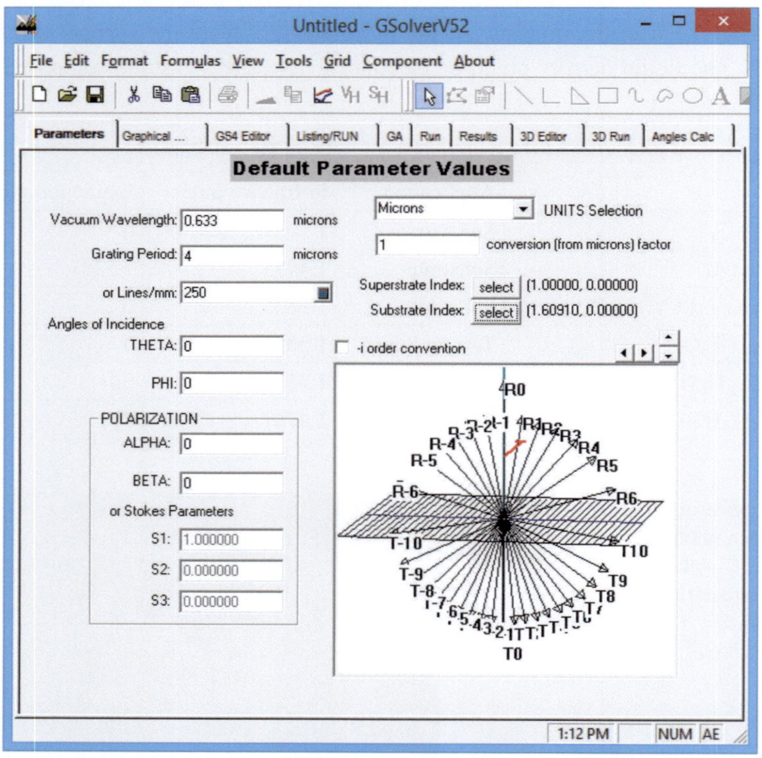

Fig. 8.34 *GSolver V5.2* software for calculation of optical parameters of microfluidics

the transmitted light were investigated with PC microstructure. Relative diffraction efficiencies of the transmitted and reflected light with three liquids and in air are presented in Figs. 8.35 and 8.36. This methodology can be applied to bioengineering.

Diffraction efficiencies of the periodic microstructures without an analyte (in air or vacuum) showed that the periodic microstructure (period 4 μm, depth 0.7 μm) was designed for operation in transmitting mode, that is, first-order diffraction maxima were three times higher in transmitting mode than in reflecting. Therefore, the periodic microstructure in transmitting mode was more sensitive to the refractive index (density) of the analyte. The refractive index changed from 1.33 (pure water) to 1.47 (glycerol) leads to the decrease of the first order maximum diffraction efficiency four times from 24 to 6%. At the same time,

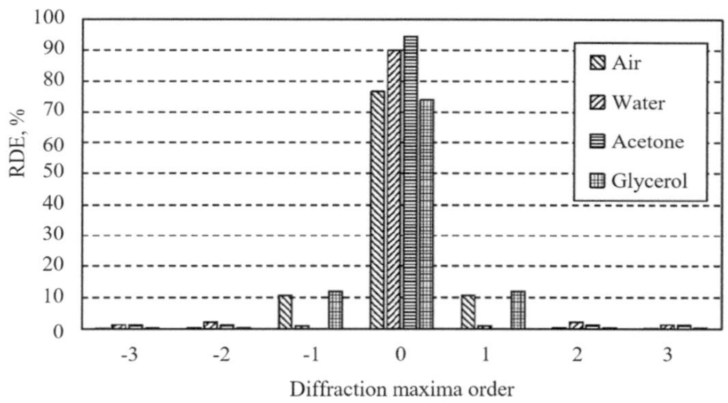

Fig. 8.35 Relative diffraction efficiencies of the light transmitted through the periodic microstructure formed in PC

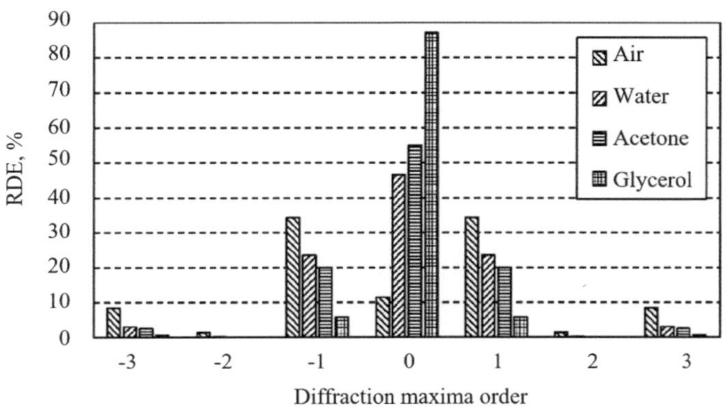

Fig. 8.36 Relative diffraction efficiencies of reflected light from the periodic microstructure formed in aluminium

the microstructure was not sensitive enough. Results confirmed the idea of liquid concentration identification in the periodic microstructures by applying numerical-experimental laser interferometric methods.

Subchapter conclusions. The refractive index and density of liquids in flow conditions could be determined from the comparison of theoretical and experimental diffraction efficiencies. Change in refractive index by 0.008 increased or decreased the diffraction efficiency of the first-order diffraction maxima by 1%. The non-uniform distribution of the fluid density could be determined using holographic methods. Harmonic variation of the fluid density could be registered using time-average or high-speed double-exposure holography.

References

1. Janusas G, Ponelyte S, Brunius A, Guobiene A, Prosycevas I, Vilkauskas A, Palevicius A (2015) Periodical microstructures based on novel piezoelectric material for biomedical applications. Sensors 15:31699–31708. https://doi.org/10.3390/s151229876
2. Bruck HA, Yang M, Kostov Y, Rasooly A (2013) Electrical percolation based biosensors. Methods 63:282–289. https://doi.org/10.1016/j.ymeth.2013.08.031
3. Ziegler C (2004) Cantilever-based biosensors. Anal Bioanal Chem 379:946–959. https://doi.org/10.1007/s00216-004-2694-y
4. Puiso J, Prosycevas I, Guobiene A, Tamulevicius S (2008) Plasmonic properties of silver in polymer. Mater Sci Eng B 149:230–236. https://doi.org/10.1016/j.mseb.2007.09.081
5. Sappia LD, Trujillo MR, Lorite I, Madrid RE, Tirado M, Comedi D, Esquinazi P (2015) Nanostructured ZnO films: a study of molecular influence on transport properties by impedance spectroscopy. Mater Sci 200:124–131. https://doi.org/10.48550/arXiv.1506.05890
6. Pérez R, Král M, Bleuler H (2012) Study of polyvinylidene fluoride (PVDF) based bimorph actuators for laser scanning actuation at kHz frequency range. Sens Actuators A 183:84–94. https://doi.org/10.1016/j.sna.2012.05.019
7. Liu H, Quan C, Tay CJ, Kobayashi T, Lee C (2011) A MEMS-based piezoelectric cantilever patterned with PZT thin film array for harvesting energy from low frequency vibrations. Phys Procedia 19:129–133. https://doi.org/10.1016/j.phpro.2011.06.136
8. Albert J, Lepinay S, Caucheteur C, DeRosa MC (2013) High resolution grating-assisted surface plasmon resonance fiber optic aptasensor. Methods 63:239–254. https://doi.org/10.1016/j.ymeth.2013.07.007
9. Tripathi SM, Bock WJ, Mikulic P, Chinnappan R, Ng A, Tolba M, Zourob M (2012) Long period grating based biosensor for the detection of *Escherichia coli* bacteria. Biosens Bioelectron 35:308–312. https://doi.org/10.1016/j.bios.2012.03.006
10. Egea AMC, Mazenq L, Trévisiol E, Paveau V, Vieu C (2013) Optical label free biodetection based on the diffraction of light by nanoscale protein gratings. Microelectron Eng 111:425–427. https://doi.org/10.1016/j.mee.2013.05.002
11. Janusas T, Pilkauskas K, Janusas G, Palevicius A (2020) Active PZT composite microfluidic channel for bioparticle manipulation. Sensors 19(9):1–15. https://doi.org/10.3390/s19092020
12. Zinin PV, Allen JS (2009) Deformation of biological cells in the acoustic field of an oscillating bubble. Phys Rev E 79(021910):1–12. https://doi.org/10.1103/PhysRevE.79.021910

13. Md Ali MA, Ostrikov KK, Khalid FA, Majlis BY, Kayani AA (2016) Active bioparticle manipulation in microfluidic systems. RSC Adv 6:113066–113094. https://doi.org/10.1039/C6RA20080J
14. Hahn P, Leibacher I, Baasch T, Dual J (2015) Numerical simulation of acoustofluidic manipulation by radiation forces and acoustic streaming for complex particles. Lab Chip 15:4302–4313. https://doi.org/10.1039/C5LC00866B
15. Pontes B, Ayala Y, Fonseca ACC, Romao LF, Amaral RF, Salgado LT, Lima FR, Farina M, Viana NB, Moura-Neto V, Nussenzveig HM (2013) Membrane elastic properties and cell function. PLOS 8(7):1–13. https://doi.org/10.1371/journal.pone.0067708
16. Petersson F, Nilsson A, Holm C, Jonsson H, Laurell T (2004) Separation of lipids from blood utilizing ultrasonic standing waves in microfluidic channels. Analyst 129(10):938–943. https://doi.org/10.1039/B409139F
17. Augustsson P, Magnusson C, Nordin M, Lilja H, Laurell T (2012) Microfluidic, label-free enrichment of prostate cancer cells in blood based on acoustophoresis. Anal Chem 84(18):7954–7962. https://doi.org/10.1021/ac301723s
18. Hammarstrom B, Laurellab T, Nilssona J (2012) Seed particle-enabled acoustic trapping of bacteria and nanoparticles in continuous flow systems. Lab Chip 12:4296–4304. https://doi.org/10.1039/C2LC40697G
19. Ding X, Lin SCS, Lapsley MI, Li S, Guo X, Chan CYK, Chiang IK, Wang L, McCoy JP, Huang TJ (2012) Standing surface acoustic wave (SSAW) based multichannel cell sorting. Lab Chip 12(21):4228–4231. https://doi.org/10.1039/C2LC40751E
20. Destgeer G, Ha BH, Park J, Jung JH, Alazzam A, Sung HJ (2015) Travelling surface acoustic waves microfluidics. Phys Procedia 70:34–37. https://doi.org/10.1016/j.phpro.2015.08.028
21. Lin SCS, Mao X, Huanga TJ (2012) Surface acoustic wave (SAW) acoustophoresis: now and beyond. Lab Chip 12(16):2766–2770. https://doi.org/10.1039/C2LC90076A
22. Reichert P, Deshmukh D, Lebovitz L, Dual J (2018) Thin film piezoelectrics for bulk acoustic wave (BAW) acoustophoresis. Lab Chip 18:3655–3667. https://doi.org/10.1039/C8LC00833G
23. Nava G, Bragheri F, Yang T, Minzioni P, Osellame R, Cristiani I, Berg-Sørensen K (2015) All-silica microfluidic optical stretcher with acoustophoretic prefocusing. Microfluid Nanofluid 19(4):837–844. https://doi.org/10.1007/s10404-015-1609-x
24. McKinstry ST, Muralt P (2004) Thin film piezoelectrics for MEMS. J Electroceram 12(1–2):7–17. https://doi.org/10.1023/B:JECR.0000033998.72845.51
25. Nama N, Barnkob R, Mao Z, Kähler CJ, Costanzo F, Huang TJ (2015) Numerical study of acoustophoretic motion of particles in a PDMS microchannel driven by surface acoustic waves. Lab Chip 15(12):2700–2709. https://doi.org/10.1039/C5LC00231A
26. Baek C, Yun JH, Wang JE, Jeong CK, Lee KJ, Park KI, Kim DK (2016) A flexible energy harvester based on a lead-free and piezoelectric BCTZ nanoparticle-polymer composite. Nanoscale 8:17632–17638. https://doi.org/10.1039/C6NR05784E
27. Belovickis J, Ivanov M, Samulionis V, Banys J, Solnyshkin A, Gavrilov SA, Nekludov KN, Shvartsman VV, Silibin MV (2018) Dielectric, ferroelectric, and piezoelectric investigation of polymer-based P(VDF-TrFE) composites. Phys Status Solidi (b) 255(3):1–6. https://doi.org/10.1002/pssb.201700196
28. Janusas G, Ponelyte S, Brunius A, Guobiene A, Vilkauskas A, Palevicius A (2016) Influence of PZT coating thickness and electrical pole alignment on microresonator properties. Sensors 16(11):1–9. https://doi.org/10.3390/s16111893
29. Zhu R, Wang Y, Zhang Z, Ma D, Wang X (2016) Synthesis of polycarbonate urethane elastomers and effects of the chemical structures on their thermal, mechanical and biocompatibility properties. Heliyon 2:e00125. https://doi.org/10.1016/j.heliyon.2016.e00125

30. Darinskii AN, Weihnacht M, Schmidt H (2017) Acoustomicrofluidic application of quasi-shear surface waves. Ultrasonics 78:10–17. https://doi.org/10.1016/j.ultras.2017.02.014
31. Guo J, Kang Y, Ai Y (2015) Radiation dominated acoustophoresis driven by surface acoustic waves. J Colloid Interface Sci 455:203–211. https://doi.org/10.1016/j.jcis.2015.05.011
32. Sazan H, Piperno S, Layani M, Magdassi S, Shpaisman H (2019) Directed assembly of nanoparticles into continuous microstructures by standing surface acoustic waves. J Colloid Interface Sci 536:701–709. https://doi.org/10.1016/j.jcis.2018.10.100
33. Zheng T, Wang C, Xu C, Hu Q, Wei S (2018) Patterning microparticles into a two-dimensional pattern using one column standing surface acoustic waves. Sens Actuators A 284:168–171. https://doi.org/10.1016/j.sna.2018.10.001
34. Zhang AL, Zha Y (2014) The breakup of digital microfluids on a piezoelectric substrate using surface acoustic waves. IEEE Trans Ultrason Ferroelectr Freq Control 61(12):2098–2105. https://doi.org/10.1109/TUFFC.2013.005997
35. Xu D, Cai F, Chen M, Li F, Wang C, Meng L, Xu D, Wang W, Wu J, Zhen H (2019) Acoustic manipulation of particles in a cylindrical cavity: theoretical and experimental study on the effects of boundary conditions. Ultrasonics 93:18–25. https://doi.org/10.1016/j.ultras.2018.10.003
36. Kumar AU, Javed A, Dubey SK (2018) Material selection for microchannel heatsink: conjugate heat transfer simulation. IOP Conf Ser Mater Sci Eng 346:1–7. https://doi.org/10.1088/1757-899X/346/1/012024
37. Xua J, Lia Y, Geb D, Liua B, Zhua M (2011) Experimental investigation on constitutive behaviour of PVB under impact loading. Int J Impact Eng 38:106–114. https://doi.org/10.1016/j.ijimpeng.2010.10.001
38. Lerch BA, Thesken JC, Bunnell CT (2018) Polymethylmethacrylate (PMMA) material test results for the capillary flow experiments (CFE). NASA/TM—2007-214835. https://ntrs.nasa.gov/search.jsp?R=20070030192
39. Pianigiani M, Kirchner R, Sovernigo E, Pozzato A, Tormen M, Schift H (2006) Effect of nanoimprint on the elastic modulus of PMMA: comparison between standard and ultrafast thermal NIL. Microelectron Eng 155:85–91. https://doi.org/10.1016/j.mee.2016.03.019
40. Ma Y, Cao X, Feng X, Ma Y, Zou H (2007) Fabrication of super-hydrophobic film from PMMA with intrinsic water contact angle below 90. Polymer 48:7455–7460. https://doi.org/10.1016/j.polymer.2007.10.038
41. Kumar P, Khan N, Kumar D (2016) Polyvinyl butyral (PVB), versatile template for designing nanocomposite/composite materials: a review. Green Chem Technol Lett 2(4):185–194. https://doi.org/10.18510/gctl.2016.244
42. Sun HB, Kawata S (2004) Two-photon photopolymerization and 3D lithographic microfabrication. Adv Polym Sci 170:169–273. https://doi.org/10.1007/b94405
43. Wu S, Serbin J, Gu M (2006) Two-photon polymerisation for three-dimensional microfabrication. J Photochem Photobiol A 181:1–11. https://doi.org/10.1016/j.jphotochem.2006.03.004
44. Stankevicius E, Gedvilas M, Voisiat B, Malinauskas M, Raciukaitis G (2013) Fabrication of periodic micro-structures by holographic lithography. Lith J Phys 53(4):227–237. https://doi.org/10.3952/physics.v53i4.2765
45. Hsieh YK, Chen SC, Huang WL, Hsu KP, Gorday KAV, Wang T, Wang J (2017) Direct micromachining of microfluidic channels on biodegradable materials using laser ablation. Polymers 9(242):1–16. https://doi.org/10.3390/polym9070242
46. Narijauskaite B, Palevicius A, Narmontas P, Ragulskis M, Janusas G (2013) High-frequency excitation for thermal imprint of microstructures into a polymer. Exp Tech 37(5):41–47. https://doi.org/10.1111/j.1747-1567.2011.00724.x

47. Attia UM, Marson S, Alcock JR (2009) Micro-injection moulding of polymer microfluidic devices. Microfluid Nanofluid 7(1):1–28. https://doi.org/10.1007/s10404-009-0421-x
48. Liu X, Ma PX (2009) Phase separation, pore structure, and properties of nanofibrous gelatin scaffolds. Biomaterials 30:4094–4103. https://doi.org/10.1016/j.biomaterials.2009.04.024
49. Salerno A, Oliviero M, Di Maio E, Iannace S, Netti P (2009) Design of porous polymeric scaffolds by gas foaming of heterogeneous blends. J Mater Sci Mater Med 20:2043–2051. https://doi.org/10.1007/s10856-009-3767-4
50. Silva DN, Gerhardt De Oliveira M, Meurer E, Meurer MI, Lopes Da Silva JV, Santa-Bárbara A (2008) Dimensional error in selective laser sintering and 3D printing of models for craniomaxillary anatomy reconstruction. J Craniomaxillofac Surg 36:443–449. https://doi.org/10.1016/j.jcms.2008.04.003
51. Mou L, Jiang X (2017) Materials for microfluidic immunoassays: a review. Adv Healthcare Mater 6(15):2017. https://doi.org/10.1002/adhm.201601403
52. Liu X, Lin B (2014) Materials used in microfluidic devices. In: Encyclopedia of microfluidics and nanofluidics. Springer, Boston, MA. https://doi.org/10.1007/978-1-4614-5491-5_859
53. Kaba AM, Jeon H, Park A, Yi K, Baek S, Park A, Kim D (2021) Cavitation-microstreaming-based lysis and DNA extraction using a laser-machined polycarbonate microfluidic chip. Sens Actuators B Chem 346:130511. https://doi.org/10.1016/j.snb.2021.130511
54. Kozuka T, Yasui K (2010) Acoustic manipulation in a microchannel. AIP Conf Proc 1474(1):363–366. https://doi.org/10.1063/1.4749370
55. Ogawa J, Kanno I, Kotera H, Wasa K, Suzuki T (2009) Development of liquid pumping devices using vibrating microchannel walls. Sens Actuators 152:211–218. https://doi.org/10.1016/j.sna.2009.04.004
56. Janušas T, Janušas G, Palevičius A (2018) Methodology for identification of liquid concentration in the periodic microstructures applying numerical-experimental laser interferometric methods. Vibroeng Procedia 19:216–220. https://doi.org/10.21595/vp.2018.20193
57. Ragulskis M, Palevicius A, Fedaravicius A, Ragulskis L (2005) Applicability of time-average fluid holography for analysis of propagating waves. Opt Eng 44(10):1–6. https://doi.org/10.1117/1.2080747
58. Tamulevicius T, Tamulevicius S, Andrulevicius M, Janusas G, Ostasevicius V, Palevicius A (2007) Optical characterization of microstructures of high aspect ratio. In: Conference on metrology, inspection, and process control for microlithography XXI, vol 6518, pp 1–9. https://doi.org/10.1117/12.714245

MIX
Papier aus verantwortungsvollen Quellen
Paper from responsible sources
FSC® C105338

If you have any concerns about our products,
you can contact us on
ProductSafety@springernature.com

In case Publisher is established outside the EU,
the EU authorized representative is:
Springer Nature Customer Service Center GmbH
Europaplatz 3, 69115 Heidelberg, Germany

Printed by Libri Plureos GmbH
in Hamburg, Germany